# COMPETING THROUGH INNOVATION

## Essential Strategies for Small and Medium-Sized Firms

Bertrand Bellon
Graham Whittington

Oak Tree Press
Dublin

Oak Tree Press
Merrion Building
Lower Merrion Street
Dublin 2, Ireland

This book is a component of a multimedia package designed by
the European Multimedia Unit of La Sept ARTE, the French arm
of the TV channel ARTE, under the responsibility of Claire
Poinsignon, with the support of the COMETT Programme and
the EUROFORM Initiative of the European Commission.
French original edition © 1994 by Ed. ECONOMICA, Paris, France
Italian edition © 1995 by FrancoAngeli s.r.l., Milano, Italy

Printed in Ireland by Colour Books Ltd.

# CONTENTS

# PREFACE

For most small and medium-sized firms, the ability to learn faster than their competitors may be their only sustainable competitive advantage.

This book has been written to enable those who are active in these firms not only to learn faster, but also to apply their learning in a systematic and strategic manner.

They are not simply large firms in miniature; nor should it be assumed that they must learn from and imitate large firms, since the opposite is increasingly the case. Small and medium-sized businesses are now the backbone of economic life and, whether directly or through sub-contracting, are involved in almost every commercial operation which takes place within the European Union. In 1995, there were over 17.5 million of these firms in Europe and they represented 71 per cent of total employment across the European Union, an average that conceals North-South differences between Sweden with 62 per cent and the UK with 64 per cent, and Italy with 83 per cent and Spain with 86 per cent. (Happily the North-South divide is not entirely uniform, Ireland having 78 per cent of its workforce in small-medium firms).

The question of the number of employees below which a business is defined as "small and medium-sized" has kept a large number of researchers in work for several decades now. But whether the preferred figure is up to 200 employees, up to 249 (the European figure from 1998) or up to 500 (and here the sector and nature of the business is the crucial factor), the process which this book describes remains the same. Throughout the book, the term "small-medium firms" is used to describe these businesses.

The challenges which this book seeks to address are common to all businesses, although individual economies have their own distinctive impact on the environment in which small-medium firms operate. For example, the UK scores relatively highly in world competitiveness studies in the areas of internationalisation (participation in international trade flows) and finance (the performance of capital markets and the quality of financial services), but rather less well when it comes to management (the extent to which businesses are managed in an innovative, profitable manner), infrastructure (the adequacy of resources and systems to serve the basic needs of business) or people, education and skills (defined as the availability and qualifications of human resources).

This relative British weakness in key areas does not necessarily mean, however, that other economies do not face the same problems. A London Business School study of best practice in Finland, Germany, the Netherlands and UK revealed the principal inhibitors of business growth to be difficulties in three areas in particular: being able to implement change quickly enough; the availability of skilled people; and international competition.

It is these obstacles to growth, which one encounters across the whole range of small-medium business activity in Europe, that this book seeks to confront.

We do not pretend that developing and sustaining "innovation as a state of mind" is an easy process. This is not a glib "how to do it" manual, but a stimulating examination, in an international context, of the issues involved in managing innovation successfully. It concentrates on the strategic and organisational skills upon which this success needs to be based.

# ACKNOWLEDGEMENTS

This book has been made possible by a European project, initiated six years ago by the European Multimedia Unit of La Sept ARTE, initially with an Italian and French team, and subsequently extended to British and Swedish partners. Many people have been involved at various stages of this project which is aimed specifically at the managers of small-medium businesses.

The book is thus the result of European collaboration and discussion involving academics, practitioners in the field of small firm development, and the managing directors and senior managers of a wealth of businesses.

As soon as the first version of each chapter was written, we worked with these partners to produce what is now a multimedia course for individual and group learning, consisting of the present volume, six films on video and a computer exercise.

The authors would like to thank Enrico Ciciotti, Simone Meyssonnier, Giovanna Monsutti, Paolo Perulli and Antoine Schoen. Particular thanks are due to Riccardo Petrella for agreeing to write the conclusion.

# *What Is Innovation?*

Innovation is change that happens in a planned manner; new processes, new procedures or new ways of organising; from the tiniest improvement to the radical rethink.

It is a specific action that is identified, discussed and agreed, which begins with an idea and ends with that idea being put into practice.

Nine out of every ten innovations are never completed; ninety-nine out of every hundred don't really change anything.

But the one which succeeds makes the other ninety-nine worth the effort.

PART ONE

# YOU DID SAY
# "INNOVATION"?

# Allegory — Episode One

The allegory which follows is not a true story but a fable imagined from a few lines written one day by the Frenchman Maurice Allais, who was to win the Nobel Prize for Economics. The allegory is in three parts. Here is the introduction. The rest of the story is to be found at the beginning of Parts 2 and 4.

## Episode One

The story begins on a remote island, inhabited by three fishermen whose sole preoccupation is obtaining enough food to ensure their survival. The only thing that they are good at is fishing by hand. Each of them works 400 hours every year to catch 200 kilos of fish, which is exactly the same as the annual consumption of his family. But these fishermen do not know how to preserve the fish by drying or salting it, and so their catch cannot be stored. The static economy established under these conditions corresponds to a total of 1,200 hours of work and 600 kilos of fish.

But supposing one of them should discover the fishing net, with the following consequences:

- The construction of this net, made from creepers that are to be found on the island, occupies one fisherman for a whole year

- This net, which requires two fishermen to use it, enables them to catch 2,400 kilos of fish each year

- A net lasts for one year.

The economic life of the island can now be based on a new economic structure, in which, in the course of each year:

- One man makes the net for the following year

- The two others fish with the net made in the course of the previous year.

This new economic structure produces an increase in output of 300 per cent over the old hand-fishing method.

This allegory highlights the concepts of productivity and savings, since during the year when they are making their own means of production, the fishermen will have to make do with the product of the activity of only two of them, thus sacrificing a part of their consumption in order to prepare for the next year. This kind of situation is well known to entrepreneurs.

Maurice Allais stopped there, but his story can be continued.

The change in their organisation of work results in the islanders catching more fish than is necessary for local consumption. So the fishermen now need to find outlets for their surplus. In this case they find them easily enough in neighbouring regions or countries on the mainland which are reasonably prosperous but lack access to the sea. In exchange, they can get cloth, tools, ropes and all kinds of implements. A new equilibrium is found, thanks to these exchanges with the outside world.

They are not, however, the only players in this game. A third region, which is also an island, is now offering fish in the market on the mainland. It offers new varieties with a more delicate taste, and these are more highly valued by the local population. Having sent observers to the fishermen's island, these competitors start using fishing nets, which they improve, and soon begin to flood the market. As prices fall, our three fishermen are on the brink of ruin.

They react quickly, however — or rather their children do, now that they have grown up and are working with their parents. They raise the stakes by developing a preservation process, transforming the fish so as to make it edible without preparation, and within a short space of time they have recaptured the market.

This respite does not last for long, however. Their invention is imitated by entrepreneurs from the third region, who are soon able to compete with them again. Yet another new riposte needs to be found.

While their fathers are running the small business with fifty employees recruited locally, the sons jet off to explore what is happening elsewhere. They return with quite an elaborate project made up of several parts which fit together a little like the pieces of a jigsaw puzzle.

They make radical changes to the fishing and operating methods. The boats are first equipped with sonar devices so as to locate the schools of fish. An electro-acoustic detection instrument identifies the kind of fish and the size of the school. The fishermen only need to lower the nets when they have the most sought-after varieties of fish in their sights.

At the same time, the brothers plan to acquire an information retrieval system which will put their business in touch with the large markets on the mainland, so that they can monitor changes in demand. Changes in the rate paid for each type of fish can therefore be responded to in real-time.

Finally, to ensure the logistics of this whole operation and to run their preparation and packaging plant in the way which reflects the market, they have to reorganise the administration unit, the rapid transportation system and the accounts department.

The very complexity of the project causes vertigo. It is no longer as simple as when they just had to increase production and sales. The traditional know-how of the fishermen is of little use to them now, and is close to becoming a hindrance. The brothers feel they are on the verge of entering a new unknown world, full of risks, with no familiar reference points and no guarantee of success. In their eyes, success or failure will depend on the roll of the dice. But given the need for speedy action, they feel that they have little choice. In the meantime, sales have fallen by 20 per cent. A family meeting is convened, and at the end of it the decision is taken to go ahead with the modernisation project as fast as possible.

But anxiety rises as night falls. Unable to contain himself any longer, one of the sons (it happens to be the second of the three), who is known to be both thoughtful and creative, goes to his father. He is less afraid of failure than he is of not knowing how to manage the project as a whole. He outlines at some length all the implementation difficulties he foresees. The father listens and remains quiet for a long time, and then questions his son about the ways in which the costs of the project will be assessed. It is now the turn of the son to sit quietly and think. He suddenly answers that the costs are well-known based on the very detailed prices provided by the supplier. This is not the problem. What worries him is how to prepare the workforce for the innovation to come. The father suggests that the first thing to do is to consult them.

The son is reassured and gets up to leave. He feels that he now holds one of the keys to the problem. From now on, he thinks, the only reason for failure can be external events which are beyond their control.

This allegory illustrates the permanent challenge which every business has to confront, as it is buffeted by changes in the external environment which destabilise an equilibrium that has been hard fought for. In the new world of trade and competitiveness in which they now have to operate, even very small businesses are forced to make strategic decisions upon which their very survival depends.

During the three decades of steady and spectacular growth which followed the Second World War, entrepreneurs in the developed economies had the opportunity to exploit many hitherto unexplored markets. It was only a question of anticipating potential consumer needs and the growth of effective demand to be able to move into mass production, creating new jobs and wealth, which in turn fed further demand. The dynamics of growth which creates its own demand could continue to be applied in markets that were not yet saturated and therefore much less competitive. Profit was based on an increase of fixed capital and a reduction of costs. This is the situation that the first part of the allegory has described.

But economies are now open to the outside. The capitalist production system now applies to the global economy. The newly industrialised countries have become formidable competitors, exploiting the advantage of their low salaries and their capacity to absorb even the most advanced technologies.

The comparative advantage of costs and prices is no longer sufficient to win an order. The conquest of markets depends on offering new products and services, on their being adapted to the ever more diversified and personalised demand, and on the added value which their use brings. Speed of reaction can certainly enable one supplier to get the better of a competitor, and has become a major new source of profit. It is a buyer's market. Innovation, in the sense of rapid and considered change, is now seen to be a decisive factor in business development, in which it has become a "strategic variable". This new reality will be the subject of the second part of the allegory.

Small firms can no longer afford to ignore technological renewal and all other forms of innovation. How can one make do without high-performance machines when most of one's competitors have invested in them? Standing still has become a much more dangerous strategy than investing heavily to move forward.

But the decision is not this simple. To innovate is to dip a toe into the whirlpool of new technologies. One innovation is often the catalyst for others, both complementary and interdependent, until a firm has assembled a real technical system. An innovation can produce a chain reaction which has an impact on the production system and on logistics, that is to say on the flow of materials, on work-in-progress, and on the products. It also has an effect on marketing, on accounts, and on financial services. In short, the new comparative advantage of a business depends on the global operational efficiency of the whole organisation.

This being the case, how does one keep control of these changes? How should one preserve the wealth of experience and know-how that is accumulated within the people in the business, and which has in many cases taken decades to develop? How is a business to remain vigilant so as to be able to seize the right opportunities at the right moment?

The size of the challenge is considerable, but the western economies have had to face up to similar challenges in the past. Today, as before, it is a question of creating a method which is appropriate for managing this change. What is at stake is the mastering of new ways of doing things, and also of a new mode of thinking creatively: competing through innovation.

*The allegory continues on page 51 . . .*

*Chapter 1*

# INNOVATION AND THE MARKET

Contents

1.1 Innovation is an Abstract Concept which Lies Hidden Behind the Activity of the Majority of Businesses

1.2 Continuity and Discontinuity in the Innovation Process

1.3 Efficiency and Effectiveness

*So many different things have been said and written about innovation that it is important to begin by clarifying the meaning of the key terms. This chapter will therefore offer a few definitions.*

*Innovation is the successful application of knowledge or techniques in new ways or for new purposes; it is also about organising a business to exploit new opportunities profitably. Innovation is for all businesses, whether they use high-technology or not. It involves an attitude of mind that is always seeking to improve, that responds to customer needs, and that aims to get ahead of the competition and stay there.*

*For a long time, innovation was felt to be limited to the period of the company's formation, or to the introduction of a new piece of equipment. But innovation has now become one of the key elements within a business, and ranks with marketing, finance, strategic planning, human resource development or production management. Innovation needs to have an impact in all these areas. Its influence cuts across the whole business.*

*Innovation is thus no longer only associated with great technological change or the success stories reported in the business pages, which are often seen as being out of reach of the typical business. Nor can innovation be reduced to the most spectacular successes, such as the French high-speed train (TGV), microchip cards, spaceships or cars which run without petrol. On the contrary, it has become an activity in which every business needs to be permanently involved, either explicitly or implicitly.*

*There are two distinct sides to innovation:*

- *A limited number of high-profile major projects, representing significant opportunities but involving substantial risks for the business*

- *A large number of micro-improvements to everyday production which increase the structural competitiveness of the business, and gradually merge with it.*

*Innovations are usually classified by types according to the degree of risk involved. They can be incremental (one step at a time) or radical; or involve, by the breadth and depth of their impact, a technological revolution.*

*Beyond the issue of uncertainty, another important element which needs to be considered involves two contradictory dimensions of innovation. The first is its discontinuity, in other words the break with the past which the business must achieve if it is to set out along a new path to growth. The second is its continuity, meaning that the innovation represents the desire to ensure that the business as a whole will enjoy a longer life than its products. The entrepreneur needs to achieve a skilful and appropriate balance between these two characteristics, making the most of the threats and the opportunities that she encounters.*

*The final element concerns the distinction between technical efficiency (rapid adoption or introduction of an innovation) and its economic effectiveness (that is to say, the extent to which economic advantages are secured by the adoption of this new process or product). The business must obviously and above all else strive for long-term effectiveness, whereas the purely technological aspect of an innovation must deliver an immediate operational advantage, in other words, be efficient.*

## 1.1 Innovation Is an Abstract Concept which Lies Hidden Behind the Activity of the Majority of Businesses

In the same way that Monsieur Jourdain (in Molière's play *Le Bourgeois Gentilhomme*) spoke in prose without realising it, most businesses innovate without being aware that they are doing so, if only because the people within them are interested in improving the efficiency of their equipment, machines, teams and services, and the quality of products, customer service and company image. The steel industry innovated when it moved on from the Bessemer converter, but reorganising the system for filing invoices, redesigning a product's packaging or reducing delivery times are all innovations as well. *An innovation is a change introduced into an economic process which has the intention and effect of enabling a more efficient use of the available resources.* Innovation thus contains four ideas:

- It consists of a voluntary act which aims to make better use of what already exists. Each individual can thus innovate at their own level.

- This voluntary act is the result of an individual seeking to satisfy personal needs and goals within a collective organisation which equally must look to achieve its own needs and goals.

- Innovation must be viewed as a challenge, and so can result in failure instead of in success.

- Where it is successful, an innovation inevitably turns into a habit or routine. So the advantage which it gives to the person or group who took the risk may only last for a limited period.

The literature on innovation offers a rich variety of definitions and classifications of the different forms which innovation assumes. This very diversity can be a source of difficulty and confusion to someone seeking practical advice which can contribute to improving the daily life of a business, or who is looking for an unambiguous overview of a complex issue. But these classifications are not without a real practical value, and certainly repay an analysis which extracts from them those elements which can be useful in practice.

It is perhaps helpful to begin with the frequently-used division of innovation into three principal types:

- **Incremental innovations** (of the small variety). These are the mass of daily improvements which each employee, or group of employees, makes to the current range of products or the current production processes. These incremental innovations accumulate almost continuously throughout the course of the production of any good or service. Obviously these improvements do not appear by chance; they occur at different rhythms depending on the period, the industry or the individual business. Although the effects of incremental innovations have an extremely important impact on productivity improvements and thus on the economy as a whole, none of them on its own has a decisive effect. But they are a reflection of the efficiency of an economy or society as a whole.

- **Radical innovations.** These are of greater scope and more specific, and they result in a total and irreversible break within a particular process: the replacement of cotton by nylon or polyethylene, the introduction of the pneumatic tyre, an organisational change, or the purchase of a machine all involve some kind of substantial break with the past. These breaks have an immediate impact not only on their immediate surroundings but also on the different elements which make up the production process. These are radical changes which do not happen in a steady stream and whose dissemination has cyclical effects.

- **Technological revolutions.** This form of innovation is due to several radical innovations occurring at the same time, and has an impact on the whole of an economy. These innovations cause the birth of new products and services, and change the cost structure and competitive position in more than just their own industry. The arrival of electrical energy or of steam power are examples of such a transformation, while the information technology explosion is perhaps the most recent example of a technological revolution. It should be emphasised that such dramatic changes are invariably followed by clusters of radical and incremental innovations which are spread by imitation throughout the whole economy. An example of this is the way in which the creation and transmission of electrical energy led to the development of small-scale electrical motors suitable for scattered businesses. But it also triggered a proliferation of domestic appliances, and led, a few decades later, to data pro cessing using computers.

This type of classification can become useful when one begins to compare and contrast the different types of innovation by degree of risk and uncertainty.

**Figure 1.1: Different Types of Innovation by Degree of Uncertainty**

---

**Total uncertainty**
- Technological revolutions

**High degree of uncertainty**
- Radical innovations (of products, procedures or processes)
- Complex R&D processes

**Medium degree of uncertainty**
- New generation of existing products

**Low uncertainty**
- Innovation by licence:
- ◊ Imitation of product innovation
- ◊ Modification of process or product
- ◊ Rapid adoption of known processes

**Virtually zero uncertainty**
- New "models":
- ◊ Product differentiation
- ◊ Slow adoption of known processes
- ◊ Minimal technical improvements

---

It will be clear from this model that innovation does not necessarily involve a technological leap which causes radical change to the way in which a business operates and which can only be judged in terms of total success or failure. Truly radical cases do exist (most of the examples come from high-technology sectors such as micro-electronics, new materials or biotechnology) but the noise which surrounds them tends to conceal the true nature of innovation, which may be less heroic and newsworthy but is invariably more closely related to everyday issues of business life.

### Figure 1.2: Types of Innovation Classified According to Products, Processes or Business Organisation

---

**Existing products**
- New raw material (e.g. a jacket made of synthetic fibre)
- Improvements in the composition of raw materials (e.g. a windscreen made of Securit glass)
- Adaptation to new customer requirements (e.g. noise level of a washing machine)
- Adaptation to new products from competitors (e.g. recyclable packaging)

**New products** (the products of some can become the processes of others)
- For new uses (e.g. miniaturisation of the audio recorder leads to a Walkman)
- For traditional uses (e.g. high frequency industrial drying)

**New processes**
- New machines (e.g. welding robot)
- New technologies (e.g. electrolytic manufacture of metals)
- New qualifications of the operators (e.g. increased multi-skilling of employees as the basis for a greater production flexibility)

**New organisations**
- New marketing and distribution structures (e.g. socks and ties on sale at rail termini)
- New ways of managing stock control, production or logistics (e.g. Benetton retail outlets supplied in real-time, due to just-in-time production)

---

According to Joseph Schumpeter's classic definition, ***innovation is the successful introduction onto the market of a new product, new process or new organisational model.***

Viewed from this angle, the act of innovating is first of all located inside the company's activity: the idea of producing and selling something new (or to produce something in a new way or to offer it in a different form) comes from the business and from its goal of developing a stronger position. It is possible on the basis of these definitions to identify different kinds of innovation according to whether the impact involves products, processes, or the ways in which production and distribution are organised.

But what actually happens is more complicated: product or process innovations are developed in parallel with innovations involving distribution, organisation or management, and these are all mutually dependent. It is by being able to appreciate and identify the range of different forms which innovation can take that an entrepreneur becomes better able to respond to (and drive) them in daily business activity.

This variety introduces the subject of the differences that exist between the concepts of discovery, invention, innovation, and research and development.

Legend has it that Archimedes cried out "Eureka" ("I've found it!") when he discovered the law of specific gravity while he was in his bath. This *discovery* was the consequence of both individual genius and the accumulated wisdom of Archimedes and his predecessors. Discovery is thus a term which covers the output of a long process and a good deal of hard work. A discovery can, however, remain in the realm of pure knowledge and science and can serve no immediate practical purpose. This is in fact what happens to most discoveries. But it can also be used to meet a specific need. Applied to a process, or a product, it becomes an *invention,* such as the steam engine or the wire used to cut cheese. Invention consists in using the imagination to create a real product. So it is an act which is both original and directly applicable at the same time.

*Innovation* is much more down-to-earth and can have an impact on all the usual activities of a business. It is always possible to make better (or at least different) use of what is available. And in a competitive environment, the sooner this occurs, the better.

Lastly, the *research and development process (R&D)* consists of systematic activity in search of those elements in a company's internal or external knowledge that can be profitably developed. One part of the research is a pursuit of knowledge for its own sake (this is pure scientific research) whereas another part aims to expand the field of practical applications (through applied technological research). The development of a prototype or of a production line introduces the theme of industrial production and a new area of problems to identify and resolve.

## 1.2 Continuity and Discontinuity in the Innovation Process

Independent of the extent of its radicalism, any innovation inevitably introduces a break in the life of a business. Understanding this consequence of innovation is essential for the success of the whole innovation process.

Alongside this break (or discontinuity), which is implicit in innovation, there is also a force for continuity, which is important to the company because it requires there to be a continual process of innovation if it is to survive the obsolescence of its products. So a business can be defined as a succession of projects, each one inevitably having only a limited life.

The product life cycle models to be found in Chapter 5 clearly illustrate this issue and its implications. They also shed light on the relationship between the length of a particular project and the lifetime of a business. From this point of view, *the innovation process is a continuous activity in the life of a business, but it is characterised by a succession of "discontinuities" relating to the introduction of particular projects*.

The capacity to guarantee the "permanent continuity of discontinuity" is thus essential if a business is to retain its place in the market and ensure its future development. The example of the French company Bic is interesting in this context: to prevent a decline caused by the maturity of the product (ball-point pens) which brought the brand fame and fortune, the company launched a number of very different products, such as disposable razors, cigarette lighters and even surf boards. The continuity of the business was thus ensured through innovative projects which led it to break with its traditional products and even sector.

Product innovation is not however the only way to secure (although never indefinitely) the future of a business. By the time a specific product or service reaches a certain level of maturity, the business needs already to have introduced process innovations in order to resist price competition. Introducing incremental innovations (such as quality improvements or different models) for existing products can also play a significant part in extending their life.

A business will therefore need, in the end, to set up a number of different innovation activities:

- Adopting process innovations to produce existing products at lower cost

- Introducing quality improvements and extending the range of models on existing products in order to prolong their life cycle

- Introducing organisational innovations within the company, independent of any specific innovation, with a view to developing an innovative culture.

*For effective innovation to occur, strategic behaviour needs to replace management, pure and simple*. This strategic behaviour is based upon:

- The ability to influence the external environment (What needs to be done in the next few years to strengthen the position of the business?), as much as a capacity to respond to it (What might happen tomorrow that will be different from today?)

- Particular ambitions concerning the capacity and the resources of the business

- A process of active management, which stimulates individuals by explaining the goals of the business (and why these are the goals) clearly, consistently and regularly; and which creates space for individual and group initiative

- Valuing human resources in a way which allows individuals to achieve their personal ambitions.

This all contributes to a construction of the future. Changes are not viewed as passive adaptations but are intentionally provoked. Forecasting replaces guessing, so as to ensure that change is not

only planned but is part of a continuous line which builds upon what has been achieved in the past.

It remains true however that the day-to-day existence of the business consists of a succession of large or small discontinuities, and that the way in which these are strategically combined determines the capacity of the business to survive.

## 1.3 Efficiency and Effectiveness

It was stated above that product, process and organisational innovations are often linked because they are integrated into the overall development of the business. But the links are in fact even closer than that. They all need simultaneously to fulfil the double criteria of efficiency (technical and productive) and effectiveness (economic, social and in terms of image).

The vast majority of innovations are adopted as a result of external stimuli: clear customer requirements or suggestions from the producer-sellers of technology; scientific publications, trade fairs, technology transfer agencies, state support, or simply the desire to copy a competitor. These stimuli tend to result in rapid changes in the technology used (the efficiency of the innovation) without the business necessarily taking all the organisational implications of this into account, or making the most of the benefits which the innovation has caused (the effectiveness of the innovation).

## Figure 1.3: Effectiveness (as Opposed to Efficiency)

- Achieving financial, market and personal goals
- Allocating resources to those areas of the business likely to produce extraordinary results
- Entrepreneurship, rather than merely administration
- Belief that the business of tomorrow must be qualitatively better than the business of today
- Doing the right things (rather than simply doing things right).

The technological choice must not be allowed to distract attention from the need clearly to define the problem that needs to be solved. Were the opposite to be the case, the danger would be that of choosing a technology in order to find what problem to solve, rather than searching for the technology that is the best adapted to the problem to solve. The Japanese solution to this problem of finding the right technology has been better than the European, as can be seen, for example, in the "rustic robots" used by the Japanese, as opposed to the highly sophisticated robots that are preferred in Europe.

Technical feasibility studies tend to show that there are usually several technological solutions able to meet the particular needs of a business. It is obvious, though — and here the distinction between effectiveness and efficiency comes into play — that there is a need for a market analysis which enables a choice to be made which fits the strategic objectives of the business. The issue of whether a technology is able to meet the needs of the business, and, still more important, whether the business is able to use it, have a critical impact on the success of the adoption process. Both elements need to happen together: the needs identified by the business are influenced by technological advances, for technology makes it possible to catch a glimpse of what future needs will be and provides new opportunities to set strategic objectives. At the same time, the search for new technologies is a necessity driven by the ambition of the business as a whole. The result of this is a positive interactive process. But if the technology is introduced into the business without being appropriate, things are likely to develop in unpredictable directions.

The *evaluation of alternatives* thus constitutes a necessary stage in the adoption process of an innovation. The business needs to consider, on the basis of an analysis of the costs and the benefits, whether it would be preferable to introduce a specific innovation rather than to continue to use traditional technologies. Although this stage marks an important point in the evaluation of the innovation project, it must be preceded by a careful evaluation of needs set against the strategic objectives which can also change under the impact of a new technology.

The example of digital control illustrates this last point. This technology which controls production operations by computer (developed particularly for machine tools but also for plastic injection presses) allows the user to overcome one of the most typical conflicts in industrial production: the choice between the rigidity of low-cost mass production and the flexibility of smaller-scale production. In reality, economies of scale are still the basic condition for commercial success, and so both options may need to be pursued; while technological progress helps the digital control technology, and more specifically the flexible manufacturing system (FMS), to allow a joint solution to be developed involving a number of models of varying complexity, depending on the level of automation required. Automation can allow flexibility and high quality on a large scale of production while still benefiting from decreasing costs. This means that a business which chooses this route can concentrate on producing the same things in a more flexible manner; but it could equally produce different things with the benefit of the same economies of scale. In this way, digital control will be even more effective as the business will use this new model not only at the level of production technology but also at the level of the business organisation, of the workshop layout, of product conception and design, of new production, and of marketing strategies.

This once more focuses attention on the question of internal resources and the relationship of the business with its external environment. In this context, external environment does not mean simply the availability of technical information relative to the new technologies. The problem is not so much one of being informed about the existence of new technologies as being trained to use them and being able to make the appropriate physical, intellectual and psychological adjustment. No entrepreneur can stay up to date with regard to all the available new productive processes, while at the same time being aware of and understanding all the strategies which could improve the market position of the business. Relations with other players in the industry can play an important role here, by helping the business not only to adopt the new technologies rapidly but also to integrate them into its own strategic objectives.

# Chapter 2

# ASSESSING THE RISK

Contents

*Small-medium businesses spend much of their time innovating, but at the same time they tend not to be particularly aware of the extent and diversity of their innovative potential. They are therefore particularly vulnerable to being pushed aside by the competition. The high level of birth and death rates of these businesses are proof both of the multiple opportunities for creation and the threat of premature failure. The disappearance of a business (bankruptcy, being taken over) can be attributed as much to the "non-success" of the introduction of an innovation, as to the "non-introduction" of innovations. In other words, faced with the challenge of technological innovation, the small firm is forced to choose between two major yet contradictory risks:*

- *The risk of innovating, which can lead to a success as much as it can lead to the total destruction of the business, or to the benefits being enjoyed by another business altogether. Medium-sized businesses, which may be in a weak market position and inadequately organised or financially insecure, often end up losing control of the results of their innovation.*

- *The risk of not innovating through adopting a purely defensive and conservative strategy. Caution leads over time to an erosion of competitiveness. The current competitive situation can easily leave a business in a state of perpetual confusion. Even if it limits its ambitions to maintaining its position in the market, a business still has to choose between all the available innovations, or better still create new product-market combinations.*

*This double risk of innovation and inactivity will be covered in detail in this chapter, viewed through the eyes of the entrepreneur needing to evaluate the shape and size of the risk before establishing her innovation strategy. It is upon the basis of this initial appraisal of the risk, supplemented by the results of the accompanying Diagnosis (a process presented in Chapters 3 and 4) that the managing director will be able to take her innovation decisions (Chapter 5). The inevitable and risky nature of change restricts the room for manoeuvre which she has at her disposal in the neverending struggle to sustain and revitalise the competitiveness of her business.*

## 2.1 The Risk Involved in Innovating

Any innovation involves a degree of risk. That much is certain. But this risk assumes a variety of forms for the huge majority of businesses which do not belong to multinational groups. Their size is the first reason for this. A 100,000 ECU investment in technology merely scratches the ICI bank balance; but for a business with fifty employees it will probably be a matter of life or death. Even beyond the size of a particular investment, the small business has little choice but to put all its eggs in the same basket. It has a specific mission, and all of its efforts need to be concentrated on achieving it; while its small size compared with a number of its competitors leave it all the more exposed to the risks involved in attempting something new. To these two characteristics must be added the usual levels of risk which innovation involves: even the best-tested techniques will spring many surprises when they are used in a new place or in a new production system. Finally, innovation carries another risk of a more social and psychological kind: an unsettled workforce, invariably torn between the anxiety of the unknown, the need to question the status quo and the excitement of novelty. The balance between these forces is a fine one to maintain.

From the start, the innovator has to confront a whole series of internal and external obstacles.

*Internal obstacles* include:

- The fact that strategic objectives refuse to sit still

- Uncertainty regarding the creation of roles and structures, as well as their stability

- The difficulty or inability to attract qualified staff

- The difficulties involved in gathering and managing information on new technology

- The difficulties associated with gathering information on competitors, their strategies and their products

- An inability to cost the innovation, even approximately.

There are also the more general difficulties involved in managing a complex situation with regard to the all-purpose nature of individuals' responsibilities (the smaller the business, the more hats the directors must wear). This complexity only serves to underline the importance of needing to have the right mindset to be able to innovate.

*External obstacles* include:

- Technological barriers

- Barriers linked to licences and standards

- Barriers due to vertical integration and to financial partners

- Collusion between competitors

- Barriers due to inexperience.

These kinds of obstacles have a habit of multiplying and accumulating. They explain high death rates among small firms and highlight their structural weakness in the face of new activities. But not everything is a handicap (or there wouldn't be any small firms at all). Their size does enable them to avoid other issues which are more likely to cause problems for large organisations:

- The stability of roles within the organisation may improve internal coherence and routine efficiency but it inhibits change

- "Group think" with regard to a perception of the problems facing the organisation, which often causes short-sightedness in the face of change

- The consolidation of privileges once they have been obtained by individuals and particularly by groups

- The maintenance of control within the hands of the founder(s), independent of any structural changes.

This over-simplified contrast of the behaviour of businesses according to their size needs, however, to be kept in proportion. Firstly, because any weakness can be turned into a strength through taking the right strategic actions, but also because the main difficulties are always due to the same key factor: the problem which everyone has in asking the searching questions which may well undermine their own position. This is a more powerful force than the fear of technical innovation. But innovation can only have an impact where it disturbs the status quo. And most people are more prepared to consider the impact of change on the security and stability of others than on themselves.

The risks associated with innovation come in three different forms:

- Commercial risk

- Intrinsic risk

- Personal risk.

**Commercial risk** appears to be the most important, although it emerges only at the end of the process, but then often in a dramatic way. ***The primary reason for introducing innovation is to strengthen the position of the business and to make a profit***. Hitting the target market (whether real or potential) is a key objective. The return on investment must be as rapid as the financial resources of the business allows. It is well-known, however, that projects tend towards optimism. This is true of any entrepreneur, but can be even more so when projects are put in place by financiers, who are remote from the realities of production and sales.

*Intrinsic risk* emerges out of the process to establish the new technology or the new organisation. The initial project must successfully overcome all the different stages of creation and implementation (with regard to delays, costs, quality, etc.). Here again, the results obtained are generally well below those initially hoped for. As far as technological innovation is concerned, the risks increase progressively, as the box below illustrates.

**Figure 2.1: The Increasing Scale of Risk**

---

1. **Improvement of existing techniques**, e.g. Airbus 300 improves each of its aircraft in the same production series

2. **Application of existing techniques with a new focus**, e.g. Airbus produces aeroplanes adaptable for cargo or passengers

3. **Applying technologies that have been tested externally**, e.g. Airbus uses the innovations developed in Concorde and copies Boeing

4. **Applying entirely new technologies**, e.g. Airbus has developed a system of electrical commands and computer assisted piloting.

---

*Personal risk*, or risk with regard to people, is linked to the idea that all the employees involved have of the project, of its likely consequences, of its success or failure. Frequently, the enthusiasm of the engineer or the promoters of the innovation triggers a counter-balancing scepticism, and a hidden or openly-voiced opposition from employees who come into only intermittent contact with the innovation.

This form of behaviour raises the question of the difference between risk and uncertainty. Lowell W. Steele (1988) expressed the risk as the mathematical product of the probability of failure (or partial success) multiplied by the scope of financial or organisational consequences. *Uncertainty, however, cannot be measured by probabilities*. It simply (but potentially devastatingly) reflects the lack of knowledge about the scope and the likelihood of the results.

For any business, there is a crucial difference between *invention* and *development*: whereas, in the case of invention, limited resources are often sufficient (the risk is thus minor), in the case of development, significant and irretrievable resources have to be committed. In many cases, individual businesses or particular inventors have a wealth of new ideas, but their development requires meetings with others (testing the ideas out) and the mobilisation of different kinds of organisational and financial resources.

For instance, James Watt invented the first steam engine in 1769, building upon the previous discoveries of Papin and Newcomen. But he only made his fortune thanks to the financier Boulton putting onto the market the first machines used in textiles, metalworking and mills. In most cases of innovation, unsuspected obstacles are revealed in the course of production and launch. As David Teece (1988) has argued, a mass of so-called innovators have discovered that technical success is certainly a necessary condition, but not a sufficient one to achieve commercial success.

## 2.2 Possible Responses to the Risk Involved in Innovating

A business is obliged to innovate, but the greater the need for and extent of change, the greater the risk it runs. Policies need therefore to be developed which limit or control these risks. These policies are varied, and they are the central concern of this book. It is, however, worth making a few general observations immediately, starting with the risk of losing control of the benefits of an innovation. This threat should lead a business towards two distinct areas of policy: protection and co-operation.

### *Protection Systems Need to be Adapted to the Nature of the Technology*

Among a number of possible positive solutions, it is best to reject as potentially suicidal the approach which consists of taking no risks at all, and therefore ignoring innovation altogether (the dodo approach, since even ostriches have a strategy).

The first response to consider is the approach which consists of making a particular effort upstream, at the decision-making level. This involves **preparing the ground**, by observing, discussing, gathering information. Here, time and energy are devoted to transforming uncertainties into risks and to reduce their number and scope. Japanese and American behaviours in this area are often contrasted. The American method consists in attempting to reach a decision as quickly as possible. The idea must become an action. The flexibility of institutions and individuals is assumed to enable all the improvement and adaptation to follow later. The Japanese, however, spend an infinite amount of time in scrupulously examining the alternatives and in interminable discussions on what appear to be only vaguely-related subjects. But this lengthy process results in a dramatic and significant decision. In this approach, the gathering and processing of the information is the key element (see Chapters 3 and 4), and may be linked to a large number of tests and practical trials.

The second response is to be found in **a judicious spreading of the risks of innovation**. A small business does not have the resources to develop a real portfolio of innovations. It can, however, for each separate innovation, contain the risk by using some or all of the following criteria: level of standardisation of the products and of the levels of complexity, or the degree of newness of the market, the product and the technology employed.

The third response is the **protection of innovation** (briefly summarised in the box below) which will be discussed in some detail in Chapter 15.

**Figure 2.2: Protecting Change**

| Nature of the Innovation | Possible Instrument for Its Protection |
|---|---|
| Of a product | Patent |
| Of a process | Copyright |
| Of internal organisations | Trade secret |
| Of image and external organisation | None |

The way to confront these risks and the levels of success which can be predicted vary from case to case. For innovations which are highly integrated into the production process, patents (which disseminate know-how) are not necessarily effective methods of protection, whereas the trade secret may well be more effective. The business can also choose to launch the product and keep the technology secret (which is the case for a number of pharmaceutical drugs and also, most famously, for Coca-Cola). Technological knowledge is sometimes more easily protected if it remains implicit.

Where product innovation is concerned, the sales and customer service capacity take precedence over legal protection. The fortune of some computer companies (even small ones) can thus depend to a large extent on their ability to create space for their own product (which may be a specialised application rather than a new computer) in an already congested market. Moreover, each business possesses, in addition to a number of strengths, significant additional assets which, where it can make the most of them, can considerably improve its position. In addition to its technological expertise, for instance, it can be excellent at knowing how to produce, or knowing how to sell, or in its understanding of closely-related technologies.

**Figure 2.3: Technological Know-how: The Complementary Skills which Make an Innovation Successful**

- Competitive production

- Efficient distribution

- Responsive after-sales service

- Positive image

- High effective demand

- Existence of complementary technologies

- Financial services (to the business and to customers)

- Ability to adopt technology rapidly.

### Using Agreements to Overcome Risk

Another possible approach to risk is not to confront it alone. The tactic of spreading risk has always existed. This section concentrates on the forms of agreement in which the innovative business retains a maximum of advantages, even though in certain cases it may depend on other agents.

**Agreements between several small firms**. This process is particularly appropriate when the small firm is involved in a production network (either local or sectoral) which has an interest in protecting product quality or defending a proven process innovation. In this way the risks of infringement, unfair competition or a lowering of quality are greatly reduced. This procedure is widespread in the Italian industrial districts. The brand is used to guarantee a respect for product quality by all producers in the association, and the protection of all finished products during the distribution process is also guaranteed, even if they are competing.

Other agreements take the form of R&D consortia. These groups bring together several partners with an interest in investing jointly in a particular phase of the production cycle, limited to research and development or to the improvement of product quality. An increasing number of examples of this form of agreement are to be found in Europe.

Using consortia can help decrease the costs of basic research, as well as reduce risk and create synergies between businesses. The question remains open as to whether the principal interest of a business in participating in a consortium is linked to the efficiency of its internal innovative process and the quality of its own technology. It remains true however that this formula is well adapted to businesses which cannot justify their own research laboratory and all the necessary equipment.

**Agreements between the small firm and a large company**. These agreements are increasingly common in an ever-growing variety of forms. *Long-term contracts* are more and more frequent, in particular in the service sector. These contracts allow companies to avoid having to bear the cost of carrying out certain tasks internally (e.g. legal, computer, accounting, scientific) at the same time as limiting transaction costs through subscription systems. Some large companies promote the development of these specialised businesses with which they are contractually linked. They can for instance guarantee the entrepreneurs a sufficient volume of business for them to be able to enter the market.

These long-term contracts may offer security to a newly-formed company, but they can also limit freedom of action and notably their scope for innovation. It has been shown that small manufacturing firms with a limited number of customers (in particular only two or three) have a lower frequency of product innovations than those which have more customers. In addition, in sectors with a history or likelihood of price wars or where technological change is frequent, long-term contracts may well include a margin for negotiation in the event of change in the market.

*Franchising* is also in rapid growth. After a period in which successes and failures (due in the first instance to the behaviour of the large companies who were placing the orders) jostled for media attention, the rules of the game have been clarified and more or less universally accepted. This allows the franchisee to enjoy a number of advantages and to avoid particular risks. Capital needs can thus be partly covered by a large "franchiser" company. At the same time, financial institutions are now more likely to lend money to a franchise than to an independent business. Franchises also benefit, due to their shared management, from economies of scale in administration and marketing.

---

### Benetton Franchising

*Benetton, a world leader in the clothing industry, developed a system which has passed in a short period through a number of organisational phases.*

*The first phase was based on the decentralisation of Benetton production into small units so as to reduce labour costs.*

*The second phase was the creation of a hierarchical network: in the centre of the group are concentrated the key strategic functions of fashion research and development, market studies and central purchasing. The centre is linked to an information retrieval service which enables production and marketing to be decentralised.*

*In the third (and current) phase, Benetton has become a multi-national company organised around the franchising of thousands of sales points in every continent.*

*Advantages: a small business has all the advantages of selling a world-famous brand name. It benefits from forecasting analysis and from market studies. It is able to benefit from the economies of scale which so large a network enjoys.*

*Disadvantages: the decisions dictated by the parent company with regard to price or to territorial restrictions could force a significant number of franchises into bankruptcy.*

---

Subcontracting is often associated with a precarious and dependent situation for the business involved. But in more and more situations these agreements guarantee a mutual advantage to the subcontractor and to the business which places the order. This is the case when the latter favours the creativity of the supplier rather than concentrating on the price of standard components.

---

### The Bosch System in Germany

*The Bosch company, based in the Baden-Württemberg region, employees 10,000 people in Stuttgart. It uses a vast array of small firms as suppliers of reliable and high-quality components. Bosch helps each business to be able to introduce a continuous stream of technological innovations, which ensures a number of advantages (in terms of quality, know-how, etc.). At the same time, Bosch's strategy (which is that of many other large businesses) is to preserve the independence of the sub-contractor: only a percentage (always below 50 per cent ) of its product will be sold to the parent company, the rest going onto the market.*

*In this way, the risks associated with excessive dependence are reduced, and the small firm is encouraged to remain open to the market, which leads to the discovery of new opportunities for innovation.*

---

***Venture capital*** involves a direct and ongoing involvement of an investment company in the activity of an innovative business (often shortly after its formation) with a view to achieving strategic advantage or to make a capital gain. For the business which brings the capital (large business or specialised bank) the investment is part of its ***technology watch*** (see Chapter 13) and can allow it to benefit from a continuous flow of new technologies developed by independent innovators. This is a flexible financial investment which interferes neither with the activity of the business which is taking the risks, nor with that of the investing company. The association remains concentrated on a specific innovation goal. However, this approach involves a high degree of risk, and is only appropriate for a small number of businesses.

---

### *Relations between Large and Small Businesses in Japan*

*Two types of relationship exist between businesses: generalised exchange and balanced exchange. In the first case, relations are driven by a set of general long-term functions. In the second case, relations are based on short specific exchanges, which involve an equilibrium between the "giving" and the "having" in each transaction.*

*In Japanese businesses, generalised exchanges tend to dominate, whereas in the West, balanced exchanges are the rule. But external relations between a business and its suppliers, customers and financiers, like the internal relations between managers and employees, have long-term characteristics. They are not merely confined to the fulfilment of contractual obligations, but require a deeper and more personal involvement from both parties.*

*The advantages are expected to be shared in the long rather than the short term; and these obligations are expressed not only in precise monetary and contractual terms, but also in emotional terms. This includes the build-up over time of mutual confidence, and the development of the ability to deal with all kinds of conflict of interest.*

*Nowadays these relations take the form in which large companies expand so as to favour the creation, by their own employees, of small independent businesses, although they sometimes make use of machines and capital supplied by the parent company in partial subcontracting relations. These new players can enjoy the benefit of the stability of contractual relations and the easier resolution of short-term problems. This does result, however, in a degree of social conformity which is not entirely compatible with the kind of risk-taking without which there is no innovation.*

---

## 2.3 The Risk of Not Innovating

Having covered the risks involved in innovation and the ways of reducing these risks, this section addresses the risk of not innovating. The threat of stagnation through inaction is a very real one for the majority of small firms. This does not deserve the label of being a deliberate strategy. It is simply the way things are.

Small-medium businesses have no shortage of reasons for remaining as they are. In the first place, they often lack up-to-date information with regard to new opportunities, or to the possibilities of adopting an innovation involving a product or process. In other cases, they lack any feel for what innovation actually is. Finally, there are cases of businesses which occupy a niche (a particular product for a particular market) which is sufficiently profitable to give the impression that there is no need to introduce new elements into its product or processes. This situation is intensified when the business sells only to its local market, or when it is in a larger market in which no substantial competitors operate.

Not innovating carries a major risk in all three cases.

In the first case, *the market can change rapidly*, and it is then essential to be ready at the right moment with the right product (at the right price).

In the second case, the *business is condemned to a gradual reduction of its technological edge* which will reduce its market share (and its share of value added).

In the third, *nothing guarantees that the niche which the business occupies will not eventually be filled by other producers*, ready to imitate or copy its products (especially if its advantage is based upon a relatively simple product resulting from a commonplace technology).

Inaction is a suicidal tactic in a world of open competition. The behaviour of competitors and customers is the principal source of vulnerability; but the business itself is a contributory factor, as it complacently witnesses, without realising what is taking place, a deterioration in relative and absolute terms of its competitive and productive position.

An illuminating example of this apparently paradoxical situation is that of the "tragedy of glass", as described by Antoine Riboud (1987).

---

### *The Tragedy of Glass*

*It is thus that the plate glass industry almost disappeared in Western Europe in the 1970s, as a result of not having spotted the importance of float glass developed by the British company, Pilkington. The engineers, accustomed to the technology of the window glass oven, invested considerable sums to carry out projects aiming to improve their processes, whereas what was happening was a real revolution. Pilkington's process made it possible to carry out, in a single operation, the two operations of production and polishing; and it supplied precisely what the market wanted: the quality of plate glass at the price of window glass.*

*There is a difficult moment, upstream from any technological project: it is when a new process must be chosen which will initially be less effective than the traditional procedure but which contains within it considerable potential to improve the quality-price ratio in the long term.*

*It is only through adopting an approach that goes beyond the immediate technical issues to consider the medium-term implications of all costs and revenue that it is possible to escape the dead hand of past technology. This dilemma can be described through representing the performance of two processes by "S" curves, as shown below.*

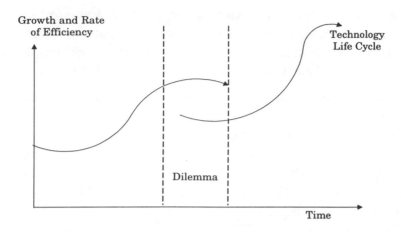

## 2.4 Possible Responses to the Risk of Not Innovating

Faced with the "status quo" syndrome, it is essential to be able to carry out a thorough evaluation of the risks involved. A few simple questions need to be asked: For how long will we be able to stay in the market with our products? Is the market expanding or contracting? What are our current and potential competitors doing? What are our human resources compared to those of our competitors? Are there any potential sources of improved productivity inside our organisation? Are any new technologies available on the market?

In all of these areas, it is essential to evaluate how secure the current situation really is and to assess the risks.

- If the business lacks information above all else, it must try to get hold of it. It may be sufficient to follow the trade press more carefully, to take part in the activities of a chamber of commerce or professional association, or to attend local, national or international trade fairs with greater enthusiasm and consistency. In other cases, especially when operating in more sophisticated sectors or markets, it is necessary to gain access to more varied sources of information.

Information (the subject is covered in depth in Part 4) is not just a question of quantity, however. In spite of even the most meticulous collection of information, the manager must *decide*

*and act from a position of imperfect knowledge*. The
quest for information is also a question of attitude: through informal daily contacts with customers, suppliers or competitors,
it is certainly possible to monitor market developments.

- If the business resists innovation because of a weakness in its
  human resources, it will be necessary to take a decision as to
  what form of professional assistance could help overcome this
  threat. In many family businesses (usually below the 150-
  employee threshold), the general management function simply
  doesn't exist, since these responsibilities are allocated to members of the owning family (though not always because they possess the appropriate skills). In this case, the addition of an external director (for example, a financial or marketing expert)
  can be a decisive investment if the business is to make the
  necessary "cultural leap". In some cases, the arrival of a new
  generation of the owning family can help revive the dynamism
  of the business (and in particular where they have been able to
  develop the necessary skills outside the business).

  The business may also contain technical weaknesses. In this
  case it can prove useful to recruit a professional with the relevant technological expertise, possibly making her a business
  partner. In other cases, policy may well need to be adapted to
  stimulate human resources (bonuses, incentives for employees
  to contribute ideas, etc.). But so complex a problem cannot be
  solved simply by selecting a formula.

- Lastly, a business can be hostile to innovation as a result of its
  confidence in its niche position and its assumed level of excellence, which may have enabled it to achieve good results in the
  past. But a business which rests on its laurels (or what it
  thinks are laurels) is in danger. To continue to occupy this
  niche, sheltered from possible competitors, demands a vigilant
  monitoring of the environment, a permanent flow of incremental innovations, an increased diversification of the product
  and a self-critical approach to the organisation as a whole. This
  amounts to *a constant state of revolution*.

# Readings and Arguments 1

This is the first of four short essays which begin by summarising a number of the key arguments developed in the corresponding section of this book, before moving on to highlight diverse approaches to the subject that are to be found in the innovation literature. Some of these approaches support the central arguments of this book; others complement it; while others represent an altogether different point of view.

The aim of these essays is to stimulate and signpost the reader who is looking for a more substantial diet than this book is able to provide.

---

### Key Arguments

- Innovation is for all businesses.

- A business is a succession of projects, each having a limited life.

- Efficiency gains are short-term; effectiveness gains are durable.

- Technology must fit both the needs of a business and its ability to exploit it.

- Where innovation is concerned, small and medium-sized firms often ignore the potential advantages that their size offers.

- The risk of innovating must always be set against the risk of doing nothing.

- The major obstacle to innovation is often a refusal to ask the really searching questions that may undermine one's own position (at least in the short term).

> - Resistance to innovation is often due to lack of information, weakness in human resources, or complacency due to past results.
> - Real innovation amounts to a constant state of revolution.

For all its value in academic terms, the debate over the nature of innovation is perhaps not immediately useful to anybody with a business to manage or a project deadline to meet. But it does reward more relaxed scrutiny.

A recent official attempt at definition highlights the breadth of the concept, which tends to ensure confusion between speaker and listener:

> Scientific and technological innovation may be considered as the transformation of an idea into a new or improved saleable product or operational process in industry and commerce or into a new approach to a social service. It thus consists of all those scientific, technological, commercial and financial steps necessary for the successful development and marketing of new or improved manufactured products, the commercial use of new or improved processes and equipment or the introduction of a new approach to a social service (Frascati Manual, 1993).

For all its apparent comprehensiveness ("Scientific, technological, commercial and financial"), this definition ignores a key area which is central to the experience of small-medium companies: the internal organisation and its ability to generate and implement change. Dosi's definition, although it is more user-friendly, clearly belongs to the "innovation has to be new" school:

> Innovation concerns the search for, and the discovery, experimentation, development, imitation, and adoption of new products, new production processes and new organisational set-ups (Dosi, 1984).

At first sight, Freeman's analysis of Schmookler's famous scissors appears to take us down the same path:

> Innovation is essentially a two-sided or coupling activity. It
> has been compared by Schmookler to the blades of a pair of
> scissors, although he himself concentrated almost entirely
> on one blade. On the one hand, it involves the recognition of
> a need or more precisely, in economic terms, a potential
> market for a new product or process. On the other hand, it
> involves technical knowledge, which may be generally
> available, but may also often include new scientific and
> technological information, the result of original research
> activity. Experimental development and design, trial pro-
> duction and marketing involve a process of "matching" the
> technical possibilities and the market . . . (Freeman, 1982).

But what he is really interested in here is the point at which the
blades meet. And this enables him to introduce a new figure, the
entrepreneur, into the discussion:

> Since technical innovation is defined by economists as the
> first commercial application or production of a new process
> or product, it follows that the crucial contribution of the en-
> trepreneur is to link the novel ideas and the market *(ibid.)*.

The entrepreneur is therefore not just a passive intermediary, but
a catalyst:

> At one extreme there may be cases where the only novelty
> lies in the idea for a new market for an existing product; at
> the other extreme, there may be cases where a new scien-
> tific discovery automatically commands a market without
> any further adaptation or development *(ibid.)*.

In these two extreme cases, the role of the entrepreneur is defined
in terms of what is exceptional about the situation. But:

> . . . the vast majority of innovations lie somewhere in be-
> tween these two extremes, and involve some imaginative
> combination of new technical possibilities and market pos-
> sibilities. Necessity may be the mother of invention, but
> procreation still requires a partner *(ibid.)*.

Here the individual has a crucial role, both as the creator, linking
"novel ideas and the market", and as controller, ensuring, for instance, that one of the scissor blades does not dominate. One-

sided innovations may require adjusting so as to take, for example, the price elasticity of the market into account.

Freeman moves on from here to make three related and significant points: firstly, that a firm which devotes resources to monitoring the changing technological environment is more likely to spot new opportunities and to exploit their potential; secondly, monitoring of the market enables a firm not only to identify what customers need but also what they are unhappy with, which may lead to product or process improvements; and thirdly:

> . . . the test of successful entrepreneurship and good management is the capacity to link together these technical and market possibilities, by combining the two flows of information *(ibid.)*.

The scissor blades have become two information flows, with the roles of entrepreneurship and management (importantly seen as two distinct functions) being to gather information, process it and then make and implement decisions based upon this analysis. This distinction between entrepreneurship and management, and the role of each in the innovation process, requires substantial development. It is not a distinction which is always apparent in the innovation literature (particularly in English, where the breadth of the term "management" appears to be an intentional source of ambiguity); nevertheless, each camp has its followers. What divides them is simple: the entrepreneurial wing focuses on a key, usually creative, individual; whereas the management lobby sees the role of management as being to harness all the resources to be found both within and outside the firm. To begin with the entrepreneur. Writing in the 1920s, Schumpeter initially describes a heroic figure who most resembles a romantic poet:

> The carrying out of new combinations is difficult and only accessible to people with certain qualities. . . . Only a few people have these qualities of leadership . . . however if one or a few have advanced with success . . . others can follow these pioneers, as they will clearly do under the stimulus of the success now attainable. Their success makes it easier for more people to follow suit, until finally the innovation becomes familiar and the acceptance of it a matter of free choice (Schumpeter, 1926).

But he goes on to tone down this trail-blazing vision with an analogy that recalls certain recent entrepreneurs:

> We do hold that entrepreneurs have an economic function as distinguished from, say, robbers. But we neither style every entrepreneur a genius or a benefactor to humanity, nor do we wish to express any opinion about the comparative merits of the social organisation in which he plays his role, or about the question whether what he does could not be effected more cheaply or efficiently in other ways *(ibid.)*.

Entrepreneurs are clearly not what they used to be; and market research would seem to bear this out. A MORI poll in 1993 found:

> . . . that only 32 per cent of Britons think entrepreneurs contribute a great deal to society (32 per cent think plumbers do too). A total of 20 per cent think that directors of large companies are big contributors to society. In Germany 60 per cent and in America 44 per cent think entrepreneurs are big contributors, way above the ratings for plumbers. And in Germany 53 per cent and in America 33 per cent give the same accolade to directors of large companies. A 1986 poll in Francois Mitterrand's France showed that 62 per cent of the public trusted private enterprise — 11 per cent more than in 1982. Maybe a Labour government is needed to convince the British that capitalism is good for them *(The Economist*, 1994).

Particularly in the wake of the so-called "enterprise society", maybe the term entrepreneur has now been so imprecisely used as to have become redundant.

Or maybe it is that the idea of the individual as sole guiding creative force has had its day. For, according to Peter Drucker (1985), who remains one of the most readable commentators on the subject, innovation is above all systematic: "Effective innovations start small. They are not grandiose. They try to do one specific thing."

This is particularly true for a small-medium firm. Unless there is clarity of thought and execution, there is nothing:

> Above all, innovation is work rather than genius. It requires knowledge. It often requires ingenuity. And it re-

quires focus. There are clearly people who are more tal-
ented as innovators than others but their talents lie in well-
defined areas. Indeed, innovators rarely work in more than
one area. For all his systematic innovative accomplish-
ments, Edison worked only in the electrical field. Any innovator in financial areas, Citibank for example, is not likely
to embark on innovations in health care (Drucker, 1985).

By working "in more than one area", therefore, Drucker means
going outside one's specialist sector. But it does not follow from
this that the ideal innovator is a specialist. For Galbraith, for instance, it is very much the opposite that is the case; and he has
the gift of conveying his argument in crisp and pithy terms:

What is innovation? How do we distinguish between invention and innovation? Invention is the creation of a new idea.
Innovation is the process of applying a new idea to create a
new process or product. Invention occurs more frequently
than innovation. In addition, the kind of innovation in
which we are interested here is the kind that becomes necessary to implement a new idea that is not consistent with
the current concept of the organisation's business. Many
new ideas that are consistent with an organisation's current
business concept are routinely generated in some companies. Those are not our current concern: here we are concerned with implementing inventions that are good ideas
but do not quite fit into the organisation's current mould.
Industry has a poor track record with this type of innovation. Most major technological changes come from outside
an industry. Mechanical typewriter manufacturers did not
introduce the electric typewriter; the electric typewriter
people did not invent the electronic typewriter; vacuum
tube companies did not introduce the transistor, and so on
. . . . (Galbraith, 1982).

This is an interesting and subtle passage, in particular in its concentration on the organisational flexibility required to implement
innovations that "do not quite fit" the current answer to the question "What business are we in?" This pushes Galbraith to develop
the idea of the "successful business innovator", whose key attribute is:

> . . . varied experience, which creates the coupling of a knowledge of means and of use in a single individual's mind. It is the generalist, not the specialist who creates an idea that differs from the firm's current business line. Specialists are inventors; generalists are innovators. . . . *(ibid.)*.

This coupling of different forms of knowledge is a long way from Freeman's information-combiner. Galbraith's vision of the experienced generalist finds its echo in Drucker (1985):

> In innovation as in any other endeavour, there is talent, there is ingenuity, and there is knowledge. But when all is said and done, what innovation requires is hard, focused, purposeful work. If diligence, persistence, and commitment are lacking, talent, ingenuity, and knowledge are of no avail.

In other words, perspiration rather than inspiration. For although Drucker accepts that there are innovations which spring from a flash of genius, he argues on numerous occasions that:

> . . . most innovations, however, especially the successful ones, result from a conscious, purposeful search for innovation opportunities which are found only in a few situations *(ibid.)*.

Significantly — and this is an insight that is of particular importance to smaller companies — he sees the most important opportunities for systematic innovation to be internal:

> Four such areas of opportunity exist within a company or industry:
> - Unexpected occurrences
> - Incongruities
> - Process needs
> - Industry and market changes.
>
> Three additional sources of opportunity exist outside a company in its social and intellectual environment:
> - Demographic changes
> - Changes in perception
> - New knowledge.
>
> True, these sources overlap, different as they may be in the nature of their risk, difficulty, and complexity, and the po-

tential for innovation may lie well in more than one area at
a time. But among them, they account for the great majority of all innovation opportunities *(ibid.)*.

Faced with these seven sources of opportunity, the typical manager, especially in a smaller organisation where the task of monitoring outside sources is disproportionately expensive, and the
likelihood of influencing them disproportionately slim, may well
be forced to admit that:

> . . . Innovation involves a fundamental element of uncer-
> tainty, which is not simply lack of all the relevant informa-
> tion about the occurrence of known events but, more fun-
> damentally, entails also the existence of techno-economic
> problems whose solution procedures are unknown; and the
> impossibility of precisely tracing consequences to actions
> (". . . if I do this, that will occur . . . " etc.) (Dosi, 1984).

Drucker's approach to the challenge is more pragmatic, and therefore closer to the situation to be found in most businesses:

> Of course innovation is risky. But so is stepping into a car
> to drive to the supermarket for a loaf of bread. All economic
> activity is by definition "high-risk". And defending yester-
> day — that is, not innovating — is far more risky than
> making tomorrow. The innovators I know are successful to
> the extent to which they define risks and confine them.
> They are successful to the extent to which they systemati-
> cally analyse the sources of innovative opportunity, then
> pinpoint the opportunity and exploit it — whether oppor-
> tunities of small and clearly-definable risk, such as exploit-
> ing the unexpected or a process need, or opportunities of
> much greater but still-definable risk, as in knowledge-based
> innovation (Drucker, 1985).

He concludes with what is at first sight just a finely-tuned paradox, but which contains a real challenge to the traditional vision
of the intuitive entrepreneur: "Successful innovators are conservative. They have to be. They are not 'risk-focused'; they are
'opportunity-focused'."

The risk, therefore, lies in not bothering to make the evaluation of alternatives (and notably the internal sources of opportunity) a habit within the business.

A different approach to the risk of not innovating arrives at similar conclusions:

> The vast majority of biological mutations are said to be harmful. When, as in human affairs, enormous numbers of random possibilities are eliminated by rational choice, the chances of harm rather than good resulting are reduced, not eliminated. The harm consists in both cases in making the individual or organisation less fit to survive in its environment than was its predecessor. Very often, the environment of the person or organisation is itself changing, so that even to maintain the same degree of fitness for survival, people and institutions may have to change their ways. So the risks attendant upon change may have to be weighed against other risks arising from maintaining the same state of affairs.

> This condition of ordinary human existence is made explicit and articulate in the institutions and procedures of industry. And in those sectors of industry in which the creation of innovations is a constant and important part of the total enterprise, the processes of change become visible in an obvious and dramatic way (Burns/Stalker, 1961).

As this passage indicates, Burns and Stalker's book, now happily back in print, is, despite its relative age, an intelligent and balanced overview, alive to the dangers inherent in the change process. But it is probably necessary to explore more recent examples of the innovation literature to encounter people-centred analysis that places innovation firmly in an organisational context. A number of these works will be discussed in the essays following Parts 2, 3 and 4 of this book. For the present, however, it is the role of the skilled manager of innovative people which concerns us:

> One of the key skills necessary for operating an innovating organisation is the skill to manage and supervise the kind of person who is likely to be an idea generator and champion — that is, people who, among other characteristics, do

not take very well to being supervised. Idea generators and champions have a great deal of ownership in their ideas. They gain their satisfaction by having "done it their way". The intrinsic satisfaction comes from the ownership and autonomy. However, idea people also need help, advice, and sounding boards. The successful sponsor soon learns how to manage these people in the same way as a producer or publisher learns how to handle the egos of their stars and writers. This style was best described by a successful sponsor:

> It's a lot like teaching your kids to ride a bike. You're there. You walk along behind. If the kid takes off, he or she never knows that they could have been helped. If they stagger a little, you lend a helping hand, undetected preferably. If they fall, you catch them. If they do something stupid, you take the bike away until they're ready.

This style is quite different from the hands-on, directive style of managers in an operating organisation. Of course, the best way to learn this style is to have been managed by it and seen it practised in an innovating organisation (Galbraith, 1982).

Or, as is perhaps more frequent, to have experienced the opposite, and to construct one's own anti-model. Nevertheless:

> . . . more than the idea generators, the sponsors need to understand the logic of innovation and to have experienced the management of innovation. Its managers need to have an intuitive feel for the task and its nuances. Managers whose only experience is in operations will not have developed the managerial style, understanding, and intuitive feel that is necessary to manage innovations because the logic of operations is counterintuitive in comparison with the logic of innovations *(ibid.)*.

Galbraith is not alone in this approach. His pragmatic focus on people is perhaps closest to what this book advocates as being the most effective approach for small-medium firms in the innovation (mine-) field.

But one crucial question remains; and it is a question which many businesses, quite rightly, ask both internal and external advocates of change. It is difficult to answer. Given the risks involved in innovating, isn't it more prudent for a "good enough" business to tread carefully and adopt a more low-risk strategy, seeing itself, for instance, as a follower rather than an innovator?

> Whatever it is that creates a difference in the characteristics of innovators and non-innovators, the consequence is likely to be that innovators are more flexible, and more able to adapt to changes in the market environment in which they operate. They clearly have the internal capabilities to respond quickly to new technological developments, and to match changing technological capabilities with changes in consumer needs. What our results suggest [an analysis was carried out into the performance of 539 UK manufacturing firms in the period 1972-1983] is that these capabilities also seem to be useful in enabling innovative firms to avoid the worst effects of recessions. If one thinks of the economic environment as a selection mechanism which "chooses" winners and rewards them with profits and growth, then the obvious question to ask is when selection pressures are likely to be at their maximum. The answer is almost certainly: "during times of adversity", and, practically speaking, this means during recessions. Most firms can survive and even prosper in a buoyant market, but only a few can prosper when the going gets tough. Innovative firms seem to belong to this second group, and that is why they outperform non-innovators.

> These results carry at least one important implication for corporate strategists. Most decision makers concerned with deciding how much R&D (and of what types) their firm ought to do tend to carry out rough cost-benefit analyses based on the view that it is the product of innovative activity which most effects the performance of firms. They project streams of incremental revenues associated with specific products and processes into the future, and compare them to cost incurred now and in the future to develop these new products and processes. If, however, the process of innovation also matters, then these calculations may seriously underestimate the total benefits of R&D activities. If a firm

becomes more perceptible, more flexible and more adaptable as a consequence of undertaking a regular stream of
research projects, then it is likely to be able to increase net
revenue from both old and new activities, and it may
discover further opportunities not yet open to it
(Geroski/Machin, 1992).

This is a significant conclusion, in an area which is often neglected. Its importance lies in its argument that "learning-by-
doing" and "learning-by-using" are the source of a significant
number of improvements and innovations:

> People and organisations, primarily firms, can learn how to
> use/improve/produce things by the very process of doing
> them, through their "informal" activities of solving produc-
> tion problems, meeting specific customers' requirements,
> overcoming various sorts of "bottlenecks", etc. (Dosi, 1984).

In other words, innovation is cumulative, and it takes time. A
business which is looking for a magic formula needs to look elsewhere.

# PART TWO

# THE INNOVATION PROCESS

# Allegory — Episode Two

*To continue the allegory which was broken off on page 7.*

The narrative was interrupted at a decisive moment in the history of the business, which, in the face of serious competition from elsewhere, can only keep up through radical modernising. The new generation of directors, the three sons of the fishermen, are on the verge of a major change of direction, as a result of a trip abroad during which they looked into new production and management techniques.

The attentive reader may recall that the middle (and worried) son and his father were in favour of organising some form of consultation with the workforce about the changes to come. But Jack, the eldest son, who has an engineering degree, takes control and, without more ado, puts into practice what he has learned to be the usual way of dealing with new arrangements.

He summons the sonar salesman and instructs him to install the detection equipment on all the boats. Within a month the whole operation is briskly completed.

A good deal longer is required, however, to set up the information retrieval system linked to the international markets. The premises are ill suited, and there are no transmission networks on the island. A specially equipped annexe has to be built, and then the connection with the cabled network on the continent has to be established, all of which involves major work and doubles the initial cost of the equipment. Since nobody in the business is qualified to operate the new machines, an experienced specialist has to be recruited whose salary is equal to that of the directors themselves. The specialist insisted on this condition before agreeing to resign from his previous job in the financial sector.

But trouble begins when the whole business needs to be reorganised in line with the fishing yield and the transmissions of the market information service. The cargo becomes more bulky but less regular. In the factory, periods of inaction alternate with days in overdrive, while the workers never know what is in store. Overtime mounts up. In spite of the efforts which everyone is making, one week there are wasted surpluses and the next missed opportunities. Turnover swings wildly between the two extremes. It is in the offices, after two or three weeks, that things begin to deteriorate rapidly. Customer complaints flood in due to mistakes in the invoices; and deliveries begin to be refused. One day, the programmer of the market information computer falls ill and has straight away to be replaced by a secretary who has only a rudimentary knowledge of the equipment. What follows is total disaster: he doesn't know how to interpret the information, and sales plunge, until the day when the system stops working altogether.

Worse is still to come as the staff in the factory go on strike, closely followed by those in the office. Within a few days, the freezers are full up and the sailors have to be laid off.

For the first time in their career, the three sons find themselves having to handle a major conflict. And coincidentally, on the same day, the bank asks them to pay the first instalment on their loan . . .

The final episode of this story is to be found on page 251. In the meantime, a few observations need to be made which will be developed over the following chapters. Technical progress does not appear to be linear. Development occurs in sequences, broken up by often abrupt changes. Innovation appears, in the short term, to break with what has happened in the past. It provokes a mixture of attitudes ranging from enthusiasm and energy in its initiators, to uncertainty and resistance among those who are subjected to them. Any acceleration in the rhythm of the introduction of new products or processes intensifies their impact and makes them appear more radical; and it heightens the importance of the human factor in the innovation process. Small businesses must (at least in theory) possess an advantage which enables them to take this into account: their size makes internal communication easier, and the availability of the directors enables them to notice the obstacles to change more rapidly. All these factors should make it easier for the workforce to come to terms with new technology.

It is clear that the engineer in the allegory underestimated this element. He overlooked the fact that people need to make sense of their actions and that the absence of explanation reinforces their resistance to change. The members of an organisation only feel part of a project when they have understood its logic and bought into its objectives. That is the precondition for their commitment and co-operation.

The management of new technologies demands, among a number of elements, new tools and new procedures. In the old system of production, chance could be kept in check by a rationalist approach based on planning and making the most of factors, calculations and probabilities. But the first experiments in innovation, in particular those involving the computerisation of procedures, showed the limitations of these traditional tools.

The course of history is scarcely predictable; uncertainty and unpredictability (and probably chaos theory as well) are now back with a vengeance in world markets. It is important in this position to preserve the ability to adapt so as to be able to ease the constraints of an over-rigid organisation of production.

We cannot however place our trust solely in simplistic rules or in intuitive navigation. The strategic management of innovations calls for a systematic intellectual rigour which is applied to *the search for information, the diagnosis of the current position of the business, the taking of strategic decisions, and the choice of techniques to adopt*. During the studies which are undertaken prior to the launch of a project, appropriate tools and evaluation methods ensure that uncertainty can be reduced to the smallest margin possible. Choosing the right criteria needs to be based on the experience of everyone involved in the business and on their collective consideration of the situation. This enables the managing director, while she remains responsible for making the final choice of strategy, to escape at last from the solitude of the long-distance runner.

What this all means is that the new approach to management involves operating simultaneously on a number of levels, and involves using rigorous pre-analysis to reduce the impact of the unpredictable, adopting a pragmatic approach to accept the risk of error, and finally demonstrating the determination and enthusiasm to involve everyone inside and close to the business in its successful development.

*The allegory concludes on page 251 . . .*

*Chapter 3*

# WHERE ARE WE NOW? —
# THE DIAGNOSIS

*This section of the book is devoted to the management of projects in which technology plays a key role. Chapters 3, 4 and 5 concentrate on the stages leading up to the decision; while Chapters 6 and 7 focus on the investment itself, and Chapters 8, 9 and 10 on managing its impact.*

*Whatever the current ambitions of the business may be, any plan for action needs to begin with a thorough examination of the current situation, or Diagnosis.*

*This involves the identification of the strengths and weaknesses of the business through an analysis of its structure, its short- and long-term performance, and its financial results. This includes identifying potential areas where immediate improvement is possible. The objective is to use this as the basis for carrying out a Technological Diagnosis. But this cannot be separated from the overall position of the business, viewed in the context of its industry, market and financial position. So it is essential that both stages are carried out.*

*The Diagnosis consists of four phases:*

1. *An assessment of the general state of the business: its history, structure, strategy and performance*

2. *An assessment of the technological position of the business and its environment*

3. *The creation of a ranking for its technological strengths and weaknesses*

4. *The identification and more detailed analysis of the opportunities.*

*The Diagnosis highlights the key factors which have an influence on the current state of the business; and it also uncovers a number of opportunities related to the use of technology and to organisational improvement which can be exploited immediately. For there lies dormant within any business a range of potential innovations which are currently ignored, but can certainly be achieved. The Diagnosis also makes it possible to assess whether the business has the potential to undertake any major new projects which would involve not only tying up capital and putting it at risk, but also possibly a gamble for the organisation as a whole. The Diagnosis also demonstrates the links which exist between the management of technology and the other areas of a business which equally demand active management. Innovation in all its forms is thus intimately tied in with the culture, values and behaviour of the business as a whole. At this stage in the process, however, the issue is not to define a new strategy or to identify future prospects; nor is this the moment to search for solutions to problems influencing the very shape of the business. The objective is solely to work at shedding light on the current situation and the factors which have led to the business being as it is, and in this it is as much interested in its strengths as in its weaknesses. Experience has shown that this phase invariably takes longer, and is more instructive through what it uncovers, than may initially appear to be the case.*

*Contrary to what one might expect, the managing director — who should after all be the person with the most accurate view of the business as a whole — should not carry out the Diagnosis alone.*

*In the first instance, she requires the active participation of all those with any responsibility within the company. Each of them possesses an intricate knowledge of one aspect of its workings, as well as a personal view of the business as a whole. Every employee should, however, have some kind of responsibility within the business. Should this not be the case, it would perhaps be useful to work with this individual's manager to try and rectify the situation.*

*The managing director should not be satisfied merely with the views of people inside the organisation, however much they may differ. It is perhaps even more necessary to have an objective view of the business from people outside it. This perspective is very important and should on no account be neglected. This additional angle in no way leads to the picture from within the organisation being disregarded; rather it enriches it through completing, contradicting and reinforcing it.*

*It is thus by taking into account three potentially very different views (that of the managing director, of insiders, and of outsiders) that a really penetrating Diagnosis can be carried out.*

## 3.1 The Diagnosis

The objective here is not to carry out a detailed financial and market analysis of the business, since this is assumed already to be an established part of business routine, involving management, the finance department, the bank and the shareholders.

The technological position, however, is intimately related to the overall state of the business; and so it is essential to know whether the business as a whole is healthy and dynamic. It is equally necessary to assess the extent to which the business is capable of taking calculated risks of the kind which were discussed in the previous chapter. This step should take the form of a structured survey, covering the full range of information produced in the course of normal business activity.

To the extent that this is not a new project, it will for the most part simply be a question of adapting it to the type of information currently available. With this in mind, the aim of this section is simply to define the range of the questions and to provide a checklist. It will obviously be necessary to adapt each question to

the nature of a specific business and industry; and also to eliminate certain indicators, while adding others. Lastly, this Diagnosis needs to include qualitative as well as quantitative information. It is not sufficient to restrict it to financial indicators. What
is significant, above all else, are climate and behaviour.

### How to Go About the Diagnosis

It is best to take a four stage approach:

1.  The structure of the business

2.  Principal strategic objectives

3.  The business environment

4.  Financial results.

**The structure of the business**. Every business has a history. It
has developed, step by step around a particular idea, a specific
individual, a precise objective, and around a variety of events of
greater or lesser significance which have become its inalienable
property. The first question which needs to be asked is this: what
elements of this history remain today?

The starting point is the observation that a business is made
up of a collection of assets, both tangible and intangible: people,
skills, investments, changes in the market, methods of finance, an
order book, public perception, etc. This is the part of the business
which is the most immediately visible, and which management
knows best. In addition, accounting legislation insists that these
assets are listed on specific dates: the balance sheet records
movement and provides a film of all the commercial and financial
transactions in the course of the last financial year. So it is very
important to go through this summary in some detail, and to
make appropriate changes to it so that it can be used as the management tool which it needs to be, particularly in the areas of the
figures for depreciation and of the risks attached to external debt-
ors.

But an analysis of the structure of a business is more complex
than this. It needs to take into account more qualitative — but
less obvious — factors: ways of doing things, ways of communicating, ways of organising. The self-image of the business is as sig-

nificant as its financial state. The reputation of its product or service is as significant as its order book.

It is these two types of information (quantitative and qualitative) which make up the checklist of questions to be found below. The Diagnosis should identify for each item both a numerical and a qualitative value. The first — and best known — questions can be rapidly covered; but the later questions, and in particular those presented in the second section of the chapter, are intended to be more demanding.

**Figure 3.1: What Is the Internal State of Health of the Business?**

---

**From the point of view of its material resources**
- Raw materials used (sourcing, price, quality, availability, potential substitutes)
- Intermediate consumption (sourcing, price, quality, availability, potential substitutes)
- Efficiency of supply and stock management, output (quantitative) and performance (qualitative)

**From the investment point of view**
- History of investment policy
- Details of investments (by kind and by purpose)
- Cost of purchase, timescale, costs of finance
- Output and efficiency of existing resources
- Maintenance requirements and costs
- Availability on the market of new equipment

**From the point of view of production costs**
- Results of analysis of management accounts
- Assessment of how the business is organised: extent of coherence, cohesion and flexibility

---

---

**From the human resource point of view**
- History of the people in the organisation and their working relationships
- Spread of roles and aptitude levels compared to the needs of the business
- Salaries (by grade and by trend)
- Profit-sharing schemes (notably with regard to innovation)
- Training (current, in-service)
- Qualifications and their appropriateness to the needs of the business

**From the point of view of financial resources**
- Current balance sheet and cash flow position
- Current cash position
- Indebtedness, cost of finance
- Potential to increase short-term or long-term capital

**From the organisational point of view**
- Applied to each of the five areas listed above

---

### Principal Strategic Objectives

This inventory can only be interpreted when compared with the long-term objectives which have existed and evolved throughout the life of the business. A strategy can be identified where these key objectives are explicit. It is also possible to distinguish between the strategy which links long-term objectives and the means to achieve them, and the day-to-day management of the business, which is by definition more short-term. Nor would it be a surprise to observe at this point the reappearance of forgotten strategic objectives, simply because they have become so embedded in the individuals within the business that they have become part of everyday life. They remain, however, essential elements in the vitality of the business.

Obviously, in this area more than elsewhere, the managers of the business are both judge and defendant, because this is what matters most to them. For this reason, it is in this area that an external opinion from an advisor can be particularly enlightening.

**Figure 3.2: Which Elements in the Diagnosis Reveal the Existence of a Strategy?**

- Long-term development goals effectively applied over the last decade
- People management; to what extent can the business be seen to be a team?
- Reputation of the business (product quality, payment record, internal relationships)
- Preliminary outline of the general issues which the business is currently facing
- Current development goals (ongoing, recently completed, planned).

These different questions have organisational (as well as marketing, financial, technological and operational) implications. There is thus a clear link between this aspect of the project and section 3.2, which concentrates on technological issues.

***The Business Environment***
A business operates and grows in an increasingly complicated environment whose impact it is necessary to assess in the form of threats and opportunities (see Figure 3.3).

**Figure 3.3: The Business Environment**

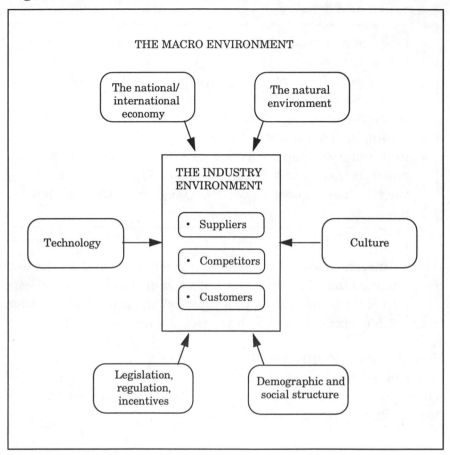

As is shown by the checklist below, a business has a broad spread of customers and suppliers (which include banks and various public bodies). Equally, a business can depend on one or several industrial groups. It also has competitors and partners. It will also be involved at some point with institutions such as the state, trade unions, local associations, pressure groups, accountants, consultants, universities, etc. Through an analysis of the nature of these partnerships, it is possible to define their particular significance and to measure their contradictions and similarities.

**Figure 3.4: What is the Environment?**

**Assessment of competitive rivalry**
- Distribution of market share
- Rate of growth
- New investment
- Known plans
- Extent of integration
- Extent and forms of competition (by price, promotion, quality, credit facilities)
- Existence or strategies of entry barriers
- Rate of new business creation
- Failure rate of businesses within the industrial sector

**Knowing the competition**
- New products
- R & D initiatives
- New contracts
- Costs and prices
- Customers
- Product improvements
- Major investments
- Changes in technology
- Diversification plans
- Collaborative activities

**Relations with suppliers**
- Cost of supplies
- Reliability of delivery times and quality
- Level of competition between suppliers
- Dependence on imports and nature of the contracts

**Relations with customers**
- Market shares, and trend
- Credit rating of major customers
- Change in demand and general outlook
- Structure of the commercial network in which the business operates

**Change in the institutional environment**
- General state of the economy
- Quality of local training infrastructure
- State of the transport system
- New regulations
- Nature of the relationship between the business and public bodies

**Sources of external advice**
- Ease of access to reliable advisors on the range of issues facing the business

### *Financial Results*

The aim of any business is to make a profit. Its success (or otherwise) in achieving this goal can be read in the end-of-year accounts. Yet everyone is aware that the accounts and financial results are only one part of the picture; it all depends on what you are looking for and how you read it. Employees will probably begin by looking at what they are getting (pay slip, benefits package, job security, quality of the job and the working environment). Management will probably also consider benefits-in-kind, the rate of growth of the business, and its reputation in the marketplace. The owners will be looking at the dividend and at the net worth of the company. The bank, the suppliers and the customers will all have their own view on how things are going. There are thus a number of possible interpretations of the same reality. Yet any analysis of the results will need to focus on three questions in particular:

- Is the business sufficiently financially sound to be able to commit itself to new ventures?

- Has the business met the expectations of each partner and participant in the past, and in particular those who will be directly or indirectly concerned in any innovation project?

- Is the business able to take further risks and to divert a part of its financial surplus into technological "adventures"; or to borrow funds while maintaining an acceptable risk?

## 3.2 The Technological Diagnosis

Every area of the business described above has a critical role to play in its activity. This is similar to having to wage a war on a number of fronts. The innovation process involves all these areas, to a greater or lesser extent. But obviously the technological section of the Diagnosis is central to the concerns of this book, and thus demands particular attention. This section will be the beginning of a process involving strategic thinking and innovative action (or even inaction, should that be the outcome of strategic thought).

It is important to recall that the term "technology" does not only cover the so-called "new technologies" and that innovation is not restricted to the use of technology. Introducing technology into a business can have a double impact: firstly, there is (it is hoped) an increase in performance compared to one's competitors at the point in time that the innovation is introduced; subsequently, downstream from this initial advantage, it is possible to benefit over a period of time from the learning curve, which may well allow the cost of production to be reduced (as long as the advantage over the competition can be maintained) as the innovation is embedded. It is thus essential to be able to measure two impacts: that of the new technology which has been brought in; and that of time, which allows the new technology to have a maximum impact, to be improved, and to be paid for.

---

### BSN and Strategic Planning

*BSN Gervais Danone is a multinational with interests in the food and glass industries. Its approach to the planning process is equally applicable in much smaller organisations. According to Antoine Riboud (1987):*

*"This is what our experience has been: every company within the BSN group draws up annually a plan covering the next three years. This is presented and discussed with the top group management. This is an opportunity to gather a maximum amount of information from outside the business, using consultants where necessary. For line managers, this is a shared moment of reflection on the strategies which will best enable them to steer their day-to-*

*day management activity. The planning cycle consists of three stages:*

- *Initial consideration of the overall objectives. This is a phase involving strategic assessment of the market, of the competition, of the business environment (from an economic, social, psychological and cultural point of view) and of the strengths and weaknesses of the business. At this stage of the process, quantifiable and accounting elements are all but banished. It is purely a qualitative exercise. It is a question of reviewing the strengths and weaknesses of the products, the needs of consumers, marketing strategies, the tactics and management of the sales force, the main areas of innovation, technological developments, the reservoirs of industrial and administrative productivity, and human resources (numbers, training, career development opportunities, reward policy). A number of strategic guidelines are defined by this process to which the line managers are committed, and against which they will be assessed over the following year.*

- *Three year plan. A few weeks later, out of these "initial objectives" will emerge a number of quantifiable proposals, which fit well together and include a detailed breakdown for each company. These details need to include product launch dates, quantities, net prices, purchase costs, direct costs, administrative overhead costs, depreciation, pay scales, profit-sharing schemes, performance targets, sources of finance, numbers involved, the creation or closure of sites, etc.*

- *Budget for Year One. The budget established in the fourth quarter of the year is a magnification of the first year of the plan, to a minute level of detail, which is negotiated with each line manager responsible for an operating unit. This will provide the performance profile for the following year, against which any deviation will be measured on a monthly basis, and where necessary corrected.*

*Our experience demonstrates the particular importance of the final stage. Without a budget, there can be no management. The stage of the measurable medium-term plan already occurs much too infrequently. But the first stage — the most important of all — is very rarely undertaken. When the point at issue is the capacity to master new technologies or to move into new markets, however, the stage of objective and qualitative strategic reflection is by far the most useful."*

---

The Technological Diagnosis is based upon six conditions being fulfilled:

1. The business already has a distinctive and identifiable "technology portfolio".

2. Management is highly familiar with a number of these technologies (but, which may come as some surprise, these will tend to be only a minority); other technologies are known only to specific individuals in the business; others are so obvious that nobody notices them any longer; others have vanished without their disappearance being observed.

3. Each of these technologies has a measurable economic weight: cost price, running costs, reliability, ease of use, fit with the other technologies which the business possesses, acceptance or otherwise by the workforce.

4. These technologies have an impact in terms of productivity and competitiveness. (What would be the cost or lost profit which the business would have to bear were any of them to disappear?)

5. These technologies are therefore different in their importance and each has its own trajectory. As a starting point for analysis, they all need to be positioned on the "technology life cycle" curve (see Chapter 5). Some technologies can be classified as emergent, others are in development, others are mature, and some are totally obsolete.

6. Comparison needs to be possible between each of these technologies and their competitors. The relative position can be measured in terms of patents, of the likely research potential which can be associated with each of these, of accumulated know-how, of equipment.

The figure below suggests a checklist which should be used to create as accurate a picture as possible of the real situation (rather than a reflection of the situation which you would like to see). The "evaluation" itself will not be carried out until the following section (3.3). Once again, the list of indicators given here is merely a suggestion, to be adapted to each individual situation. At the same time, every element which is identified needs to be given a relative weighting in a five-year time-frame and in relation to the competition. Equally, it must be set against objectives which are already in place and the available resources for implementation. Here (once again) the listing of technologies is only of interest as a trigger to further questions about the potential which currently exists. It can best be described *as an exercise designed to destabilise*, through asking searching questions as to how healthy the current situation really is.

**Figure 3.5: Technological Assets**

---

**Technological inventory of the business**
- Products, registered trademarks, patents
- Models, design

**Manufacturing know-how**
- Know-how of each process
- Products, their potential and rate of improvement and of differentiation
- Non-patented production technologies: conception, materials, production, assembly, quality control, marketing, etc.
- Control over analysis and influence on quality/cost
- Add to this the known flows currently coming from outside (see Chapter 13 on technology watch)
- Processes and equipment
- Potential markets and possible substitute markets

---

**Marketing know-how**
- Knowledge of competitor behaviour
- Knowledge of customer behaviour and needs
- Organisation of sales office
- Organisation of after-sales services

**People in the business**
- Knowledge base of technicians, engineers, researchers
- Levels of qualification (spread and development)
- Potential of workforce, and attitude towards training and towards change
- Team spirit, level of integration and spread of best practice inside the organisation

Once it has been put into some kind of order, this first level of analysis enables the true strengths of the business to emerge clearly. But it does not necessarily identify the way in which they are exploited. So it is now necessary to go further and to draw out the key aspects of the technological potential of the business in terms of what the market requires. This will result in us being able to come up with a *strategic matrix of the key know-how of the business* through a classification of the different elements of the firm's tangible and intangible assets. The choice of which technologies to keep is obviously not an easy one: it is necessary to be able to identify them, and to be confident in their role as drivers compared to the other technologies which the organisation uses. It is also necessary to be able to assess the likely extent of the firm's control over the future development of this technology. This highlights the need for the checklist of technological assets to be carried out in two stages: firstly, identifying them; and then examining the relationship between them.

### 3.3 Ranking the Strengths and Weaknesses

The first stage of the process has now been described: a photograph of the business in the form of a reasoned analysis. It consisted of identifying, listing and then double-checking that nothing significant has been ignored in an area of the business which is often not sufficiently familiar to management. The process of ranking and comparing the business with its competitors represents the beginning of the establishment of an order of priority. This stage of the process can be carried out by one or more employees, without necessarily needing to mobilise the organisation as a whole. From this point on, the exercise will become increasingly collective and more dynamic. Establishing an order clearly implies a series of value judgements, and a genuine examination of the different individuals within the business and of how they work together. This self-questioning demands clear thought, but also a genuine involvement of these players with the goals of the organisation as a whole. Central to this stage of the process should be the creation of a Consultation Group, which will be more fully dealt with in the following chapter.

It is now necessary to examine in greater detail the observations which have just been made. It is proposed to do this in two straightforward ways: firstly, to establish a ranking of the technological strengths and weaknesses; and secondly, to identify the opportunities. This process will not lead directly to one of those large strategic matrices beloved of consultants, but the experience of most of the firms which carry this out suggests that this type of ranking, where it is the result of real consultation between the people in the business, can produce remarkable operational outcomes.

The first step should be to list in order of importance the technological strengths and weaknesses of the business. Such a selection needs to be carried out in a realistic manner. In later reflection on innovation projects, it is certain that making things happen will be accompanied by a number of unexpected difficulties. Diagnosis is thus first and foremost a critical exercise which must bring into the open a range of difficulties which have perhaps escaped identification at an earlier stage. This does not mean, however, that optimism is not allowed.

We also know that what initially appear as strengths can turn out to be weaknesses if their context changes; and the opposite is equally true. Other factors may well remain in the unpredictable state of being classed as *dilemmas*. Finally, we know that certain existing technologies cannot realistically be viewed either as strengths or as weaknesses (for instance, when everyone else has them as well). They are thus external factors, or relatively neutral elements, to which the business needs to adapt, but without being able to change or make use of them to its own sole advantage.

## Figure 3.6: Results of a Diagnosis

*Output of a ranking of the strengths and weaknesses of a street furniture and playground equipment business (with 10 employees) after 13 years of trading:*

**"The strengths of our business"**

1. First of all, design. A strong product concept, designed to meet the objectives of solidity and classical line, to be used in large public open spaces. Simple but entirely reliable technologies (materials, structures, cogs, fixing devices). The products are more solid than those of our competitors. A "Meccano" feel to the product. At its launch, the first of its kind in its assembly method and the material used.

2. Good reputation with its customer base. Thirteen years on, the materials and structures are still fit for the purpose even in situations of heavy usage (city, suburbs).

3. A growing and creditworthy market. The market was not particularly competitive at the outset.

4. All products are first presented to the customer in the form of a 1-to-10-scale kit. This aspect of technical back-up to the sales team is offered to the customer at no charge, and can be kept whether a contract is signed or not (the proposal can go in the bin, but the kit remains on the customer's desk).

5. Input from outside technicians due to an annual placement involving an engineering student from the local university (although there are problems attached to evaluating results and integration into the process as a whole).

To the above list, it would have been possible to add the following three points:

6. A small but reasonably effective well-balanced team. High quality of work and speed off the mark.

7. A business which is still going strong after thirteen years. Although not large, it has a good image among customers.

8. From the commercial point of view, the catalogue is good (but the lack of new products is beginning to become a weakness).

**"The weaknesses of our business"**

1. Our competitors have developed more flexible assembly systems: they are able to change the shape of their structures; while our playground system cannot be adapted to make a round shape (our materials are square); nor can we introduce colours.

2. We have specifically targeted the large open-spaces market for which our material is appropriate. Our product is ill-adapted to the materials required for schools and similar institutions. Gaps at the level of product diversification.

3. We have attempted to dress our products on the "Barbie Doll" principle: the structures are sound and proven, but you need to dress them in coloured accessories. We are less good in this area, through a lack of imagination. Our mind-set seems ill-suited to resolving simple technical problems.

4. Day-to-day concerns tend to overwhelm the innovative effort which is necessary if the initial creation is to be updated and developed. The managing director is still behaving as if he were running the company as it was when it was only one-third of its current size. The theoretical improvements which he comes up with never come to fruition. He does not really know how to delegate, and employees cannot take an initiative without asking "Mr. Mark" (particularly in the design department).

5. Even after thirteen years of trading, the company has still not managed to build up a bank balance sufficiently large to be able to handle the lengthy decision-making and payment delays associated with public bodies.

To the above could have been added:

6. A growing but very unbalanced distribution network, with insufficient attention being given to the most competitive market segment in particular.

7. Inadequate general training being given to staff (e.g. computer failures).

8. A sales-led obsession with developing turnover at all costs; little real control over margins and costs. The sales network is equally vulnerable to this criticism. Overheads always seem to be rising faster than turnover.

9. Poor stock organisation and control leads to delays in responding to any variety among orders.

10. Marginal product defects which our claim of perfection makes even more noticeable.

11. Insufficient control of the subcontractors who are responsible for the essential task of producing the pieces (too much acceptance of "not quite good enough"); a lack of professionalism (terms of reference), while the business and the subcontractors develop and grow. Loss of control both of costs and quality.

12. The geographical position of the business is a real problem, since it is a long way from major conurbations; this is less of a problem for transport costs than for a strong sense of being out on a limb. Even more than to our subcontractors, we find it difficult to get close to the market.

---

**"Uncertainties"**

1. The market is growing more rapidly than it did in the past, and the competition has grown, over the last ten years, from 20 to 150. Although it is growing more slowly, demand has become more critical where quality is concerned.

2. Vulnerable when faced with variations in the order book (due to there being only one sort of product and only one sort of customer: local communities). The market is directly related to the budgets of the local administrations and to political factors. There is also the problem of barren periods due to seasonal budgets and municipal elections, or to the impact of the national and international economic situation.

3. An opening-up of the market which is difficult to control. Are we able to identify changing customer requirements in sufficient time to be able to respond to them?

---

How should a business go about ranking technological strengths and weaknesses?

Firstly, they can be seen in terms of their commercial impact: they can be measured by their real or potential impact on the market. This judgement obviously needs to take the time factor into account. A major innovation push lasting four years may lead to an irrecoverable financial deficit but, in the medium term, become a strength through opening up the route to diversification. The weaknesses reveal the bottlenecks which can undermine the potential for growth in other areas of the business.

These factors all need to be evaluated against a sufficiently long time-frame (at least five years), through drawing out the major changes and their causes; these apply equally to established techniques and products as they do to those which are more recent or even those that are still in the process of development.

Finally, they must be isolated, reduced to a maximum of six, and ranked in order of importance. The establishment of this list and the decisions which are made as to its order are crucial, and their accuracy and usefulness will reflect the quality and depth of analysis of the preceding discussions. Once this work has been carried out, it becomes necessary to relate these strengths and weaknesses to each other and to define precisely their influence on each other: to what extent is a particular weakness compensated for by a specific strength? But to what extent, on the other hand, does a weakness reduce the beneficial impact of a particular strength?

**Figure 3.7: Strengths and Weaknesses Applied to the Example Above**

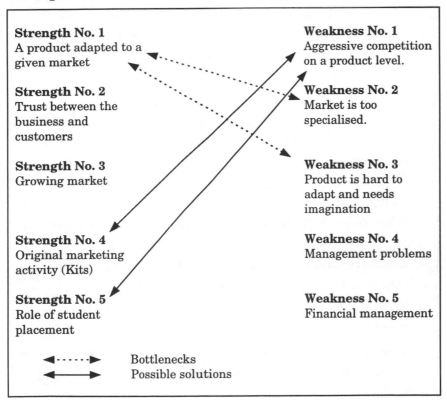

## 3.4 Identifying Opportunities

By this stage of the process, the key technological characteristics of the business will have been identified; as will its current strengths and weaknesses concerning the organisation of its resources, processes and products; and also the extent of their integration with the other elements which make up the business. This analysis is very much the result of a collaboration which involves a broad range of actors within the organisation (see the following chapter). The internal Diagnosis will have been set against an external one, and the findings thoroughly discussed, so as to arrive at a consensus involving the greatest possible number of employees.

But it is not yet the moment to talk about strategy or the taking of major decisions. For such a detailed photograph, assembled and discussed by theme, contains a depth of field and richness of colour which must be exploited at all costs. Without anticipating later stages in the process of arriving at a decision, a number of immediate opportunities can be identified through uncovering the layers of improved productivity, efficiency and innovation which already exist and are clearly within reach.

**Figure 3.8: Identifying Immediate Opportunities**

- In what areas can action immediately take place within the business?
- What are the external opportunities which should immediately be grasped?
- What scope is there to improve the market position of the business at little cost?
- Are there any other applications for an existing product?
- Can features be added to an existing product?
- What would result if some components were deleted?
- What other products are made in similar ways?
- What products look similar? Can any ideas be copied?
- What materials can be substituted in the product?
- Is there a better alternative to any of the process steps?
- Is it possible to sell another product to current customers?
- Who are the potential customers of the product?
- Are there other ways to market the product?

This process and its outcomes are in no way a substitute for the time and thought which need to be devoted to the medium and longer terms (and which will be described in the following chapters), but it is very much a starting point of this process.

There are two other sources from which opportunities can emerge which need to be added to the observations to be found above:

**Figure 3.9: Internal and External Opportunities for Uncovering Dormant Potential Innovation**

---

**Internal**

- Better use of existing equipment
- New use of existing unused technologies
- Workforce either under-employed or working beyond current skills level
- Projects started but, for a variety of reasons, never followed through
- Prior decisions never implemented
- Specific training never fully exploited

**External**

- Results of technology watch (see Chapter 13), existence of available patents in the same area
- Proposals already made or immediately feasible for co-operation with other firms on specific innovation projects
- Customer suggestions for immediate improvements

---

These opportunities emerge almost naturally from a thorough Diagnosis. They do not revolutionise the life of the business, nor do they force it to adapt its strategic objectives. Their impact is rather to give back a little flexibility to the organisation as a whole, as a result of which the commercial position of the business is likely to improve.

# Chapter 4

# THE CONSULTATION GROUP

---

*Contents*

---

*The need for a Diagnosis of the business was introduced in the previous chapter, but without specifying who should be responsible for it. This is a much more important and far-reaching question than it perhaps appears to be, and demands a whole chapter of its own. This chapter describes the issues and the limitations of the different stages of a process which need to become an essential part of life in any business wanting to achieve success. Before moving from the Diagnosis to the stage at which a decision is made concerning a specific technological project (launch of a new product range, introduction of a new production process, or computerisation of stock control, personnel or customer control data) it is necessary to resolve this question of organisation; for it is central to the whole approach which this book describes.*

*It has become a truism to stress the increasing complexity of business life, the inevitable constraints on all types of cost, the importance of product quality and after-sales service, or the key role played by professionals and specialists in each area of the business. The universal man or woman, capable of combining all these qualities, has never existed and can even bring about his or her own downfall through pretending to be so in everyday life.*

*Innovation is precisely the area of business life where one en-*
*counters the greatest complexity, need for specialisation and risk-*
*taking. Before any and every major decision, the allocation of roles*
*needs to take place, involving consultation and consensus, and*
*taking into account the points of view of everyone with a stake in*
*the future of the business; all of whom will be affected, in one way*
*or another, by the projected changes. It is therefore necessary to put*
*in place what we will call a Consultation Group, although some*
*users may prefer the terms of task group or working group, de-*
*pending on the nature and culture of the business.*

*No one is pretending that the proposals which follow are easy to*
*implement; but then current trading conditions do not simply al-*
*low a business to put a lick of paint on the traditional ways of do-*
*ing things.*

*The starting point in this process is the observation that the key*
*manager in a business is no longer the sole decision-maker in a*
*strategic technological and organisational project. This is not*
*through incapacity nor does it remove the likelihood that the final*
*decision will probably be taken by this key manager alone (see*
*Chapter 5). But experience has shown that strategic innovations*
*are events which directly involve — as individuals — too many*
*people in the business to be brought safely home to harbour using*
*the traditional autocratic decision-making processes. Even the*
*most "technical" or "commercial" innovations can profoundly*
*shake up working methods and the image which employees have*
*both of themselves and of the business as a whole. Success will*
*prove elusive unless everyone in the organisation is committed to a*
*project that has been initiated at the top, and management needs*
*to understand and buy into the employee point of view. Dialogue is*
*therefore not a luxury, but an absolute necessity.*

*The consultation process consists of wide-ranging discussion*
*and exchange of views prior to strategic decisions being taken; but*
*at the same time the structure needs to be kept to a minimum so as*
*to be certain that uninhibited questioning and debate occurs at*
*appropriate points in the whole process. The idea of a Consultation*
*Group thus requires the development of a flexible framework*
*which is able to stimulate discussion on a clearly-defined subject.*

*Each stage of a technological consultation consists of five elements: analysis, debate, proposal, follow-up, evaluation.*

*It will be clear from this list that neither the final decision nor its implementation should take place at this stage of the consultation process. The consultation framework needs to be capable of analysing and discussing a situation, of putting forward ideas and then of advancing a range of proposals concerning the long-term direction of one or more projects. The final decision must however fall to the "government" of the business, which needs both to make the decision and take the lead in directing its implementation.*

*The Consultation Group will thus have a central role from the beginning to the end of the innovation process; from the initial germ of an idea to the evaluation of its eventual impact. This will probably require there to be a change of behaviour within the business, which needs to become flexible and able to change shape. The object of this approach is to enable different points of view to be taken into account as and when issues emerge. What it is definitely not is a way of creating a new layer of bureaucracy.*

*It should already be clear that the Consultation Group will be assisted in its role by a number of guiding principles. In particular, it needs:*

- *To reflect the broadest range of points of view on the subject to be dealt with*

- *To operate with precise objectives with regard to the issues at stake and to timescales*

- *To enjoy legitimacy throughout the organisation, and sufficient resources to be able to carry out its task*

- *To have considerable freedom in which to operate*

- *To be positioned so as to be able to influence the decision-making process.*

**Figure 4.1: Opportunity to Innovate Internally and Externally**

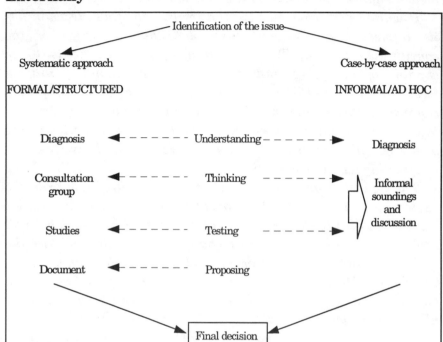

## 4.1  The First Question: Who Will Carry Out the Diagnosis?

It will be clear from the preceding pages that carrying out a Diagnosis is not merely a preliminary to the formulation of strategy. It is already a strategic act in itself.

The issue of what form this Diagnosis should take and how it should be carried forward is thus crucial.

The managing director cannot carry out a Diagnosis alone: firstly, she is not in command of the full range of information required (even though she may be convinced that this is the case). In addition, she cannot be both judge and defendant, given that she will almost inevitably favour her own story. When a doctor is concerned about her own state of health, she goes to see a colleague. In the same way it is entirely reasonable that a managing director should seek assistance. This support consists initially of everyone inside the business, and subsequently (where this is appropriate) of a fresh outside view. The Diagnosis is thus very much a group activity.

The previous chapter described the importance of using the Diagnosis as a management tool in the running of the business. It can also be regarded, for a number of reasons, as the first stage in the consultation process: the Diagnosis must be based upon all the information that is available at a given moment; in this way, it benefits from different points of view, but equally it already sets in motion the process by which the employees who participate in it begin to modify their own attitudes to the idea of change and to feel greater involvement in the objectives of the whole organisation. The consultation process should not however be confused (as too easily happens) with a meeting to which staff come to hear what their managers have already decided.

Each employee has a personal experience of the recent past, and this experience is always richer than it perhaps appears to be to the individual concerned; and each employee will doubtless in the past have had numerous opportunities to discuss and contrast this experience with colleagues, or in small groups. He is thus able to make a contribution to the Diagnosis, even if it is only to confirm an observation or point of view that has already been made elsewhere. This is the easiest, fastest and most efficient way for some kind of internal consensus to develop. It is also through this approach that process innovations are identified. These can be put into practice almost immediately and have equally immediate effects (however insignificant each one may seem to be) on the overall position of the business.

But who should be in the group which will carry out the Diagnosis?

1. The first group who need to be involved are those employees whose role includes generating and collecting information: the accountant, the human resources manager, and the people responsible for control of administration, sales and stock.

2. *But the mere accumulation of specialist information does not in itself produce a well-organised and useful Diagnosis*. So the first attempt to give some kind of shape to the mass of information that is gathered probably needs to be the work of a single person. It is advisable for this task to be

carried out by a member of staff with a neutral (or generalist's) point of view, so as to reduce the risk of bias, and to ensure that no single source of information is ignored.

3. The next stage is to be able to ask the right questions. One option here is to add to the team *someone from outside the business* who can both ask objective questions and provide immediate expert advice. The process of defining the boundaries of the Diagnosis can be achieved in a number of ways, but must never be seen as being only of minor importance, or as being isolated from what comes before and what follows. The best solution is probably that, once the initial information gathering has been completed (but before the staff meeting), the outside expert should work directly with the manager who has initiated the Diagnosis, with the member of staff responsible for assembling the information and with the Consultation Group. It is important to involve this outsider in the daily life of the business (as will already be the case with the firm's accountant and legal adviser), but the difficulties associated with this should not be ignored.

4. Once the data and preliminary observations (which should take the form of very short written documents that are widely circulated) have been collected and put into some kind of order, these can be examined by the Consultation Group. This group should include, where necessary, the managing director, the head of sales, the systems manager, those people who have direct customer contact, and the engineers and technicians. Participation in this group needs to cut across the whole organisation. It is important that anybody whose role in the business leads them to have an opinion on the subject should be able to participate.

To assemble a Consultation Group is not as difficult a task as it may at first appear to be. Experience shows that a flexible and well thought-out framework leads, in almost all cases, to the creation of a consensus around the evaluation (which itself already includes a number of innovative features).

But experience also shows that the Diagnosis often exposes the inadequacies of the information systems within the business itself, both with regard to the amount that they generate and to their relevance to what is really needed. In many cases, therefore, it may be decided that it is necessary *to construct some kind of new or renewed customised technological information system*. There is no doubt that it will become clear, by the end of the process, what information is really necessary; and existing systems can then be adapted so as to monitor the business and its development more effectively.

## 4.2 Consultation Shouldn't Be Restricted to the Diagnosis

The Technological Diagnosis is only the tip of the iceberg which is the management of innovation within the business. The process can certainly have a permanent effect on internal behaviour, and employee ownership of it is part of a strategic rethink which can restore the overall sense of purpose of the business. This in turn can become the starting point for serious consideration of the different strategic options, and of the various approaches which can be adopted towards achieving these objectives.

The outcome of all this is therefore that the aims of the Diagnosis have now been somewhat extended. It has developed into a project for the whole business; and so, at this point, *the technological projects begin to be seen as one aspect of a more general objective setting* (which is, after all, what they should have never stopped being in the first place).

### The Business As a Focus for Profit, Assets, Values and Aspirations

Any business exists to make a profit. But other factors always come into play, and *the narrow pursuit of profit often leads to a business being less profitable than it might otherwise have been*. For every business is extremely complex. It is built around a number of assets, which have already been referred to in the previous chapter in the context of the checklist of resources:

- Capital

- Buildings, machinery and vehicles

- Intangible assets: patents, reputation, advertising

- Potential customers

- Close customer relationships

- Information

- Economic dynamism of the area

- People, their skills, their qualifications and their experience

- Image, based upon internal and external perception of the organisation and its output

- The complex and continually evolving network of relationships which bind all other elements together.

The last three points above show the deeper foundations upon which the true values of any business are built. These values are more likely to be implicit in everyday activity rather than explicitly stated objectives. They most frequently manifest themselves in the following ways:

- Product quality

- Past reputation of the business and its leaders

- Quality and permanence of its internal relationships

- Confidence in the business from suppliers and customers.

In addition, the business also represents an aspiration, or more usually *a series of aspirations* which do not always coincide:

- Those of the management team

- Those of each member of staff

- Those of every partner of the business: suppliers, customers, competitors.

These aspirations may be clearly visible or they may be hidden, but they certainly exist. Every individual is always looking to add something to his life, and it is hardly surprising that some of this extra should be sought in the workplace.

These issues are very closely linked to the subject under discussion. For technological innovation, like any major project, depends for its success on taking these different factors, values and aspirations into account.

### A Broader and More Active Approach to Consultation

So it is entirely logical that this consultation process should not be restricted to the current situation alone, but will probably need to be extended so as to include not only discussion of long-term objectives, but also to cover the other factors which have an influence on the nature and extent of innovative behaviour.

Obviously the individuals with greatest interest in the consultation process will be those who see themselves as the producers of innovation. The innovator is an enthusiast: whether involved in discovering the potential uses for information technology, coming up with a design for a new product, or rationalising existing production methods, he takes a delight in making things happen. To achieve this, he will often be focused exclusively on a single goal, working intensively way beyond normal working hours, and bringing a perfectionist approach to his work.

At the same time, he will often resist being dictated to by any decision which opposes or invalidates either the systems that he has put in place or the machine or organisational change which he considers to be more effective. He is able quickly to get to the heart of the decision-making process and wants to be involved in the choice of materials, of expenditure or of production processes. The consultation process enables him (for better or for worse) to get a fair hearing.

This creativity and enthusiasm can lead to a greater sharing of knowledge, and also (if maximum benefit is to be achieved) to a greater sharing of power. Participation in decisions over expenditure, for example, can create a stronger desire to participate at the design stage, and this in turn implies greater involvement in the setting up of new organisational arrangements. These closer working relationships and higher levels of involvement occur across the organisation as well as vertically. By placing individuals on a more equal footing, it can lead to their developing a greater respect for each other based upon their abilities rather than simply upon their place in the hierarchy.

### *The Terms of Reference of the Consultation Group*

On this level, the Consultation Group can function very much as a testing ground for the strategic decisions of the business. It is responsible for proposing tactics and strategies. These strategies are very much the tools which assist in the decision-making process, and so they need to be sufficiently varied to provide real alternatives for a final decision. But it should not be assumed that a decision to innovate has to be made or implemented immediately. It needs to be tested and compared with other options before a final decision can be taken. It should be clear, however, that once a final decision is made, it will be difficult, and probably impossible, to turn back. Each of the strategic options must therefore be described and analysed in a number of different ways. Here is a list of five of these, including details of the indicators against which they need to be evaluated:

**The strategic dimension:** to what extent does such a strategy have the potential to become the guiding light of the business and its employees for years to come? To what extent is this new product line likely to become one of the key areas of activity and to be seen as such by the customers?

The factors to be taken into account are the following:

- Extent to which the key information required can be identified, gathered, analysed and acted upon at the different levels of the business

- Image that the business wants to convey or maintain among its customer base

- Potential interest of the employees

- Expectation of development in technologies used and in markets.

**The technological dimension (including production):** it is necessary to define precisely what is covered by the new technological dimension and its feasibility. This should be in relation both to the current asset base, and to what is carried out elsewhere and can be acquired or copied.

Factors which need to be taken into account:

- Changes in productivity
- Changes in production processes
- Changes in process organisation
- Knock-on effect on any other area (e.g. administration)
- Knock-on effect on other areas of production.

**The marketing dimension:** 50 per cent of industrial innovations are born out of the need to meet customer demands and potential needs. If a business is able to respond to a specific requirement in reasonable time and at an acceptable price, an innovation has every chance of success.

Factors which need to be taken into account:

- The formulation of customer needs
- The extent to which the need is going to be an ongoing one
- The reality of the need
- An evaluation of competitive behaviour, both on a national level and abroad.

**The financial dimension:** if a strategy doesn't stand a good chance of opening up a new area of activity within a realistic period of time (taking into account the likely financial return against the cost of obtaining finance), it is pointless to take the risk.

Factors which need to be taken into account:

- An analysis of the costs of the project
- A calculation of the risks involved
- The current financial capacity of the business.

**A cultural dimension:** any major strategic project has got to be able to carry people along with it. It is only through a shared enthusiasm that it will be possible to mobilise the energy and resources of a business over a relatively long period.

Factors which need to be taken into account:

- Internal profile of the project

- External profile of the project

- The enthusiasm and commitment on the part of suppliers and customers with a role to play in the process of change

- The extent to which different types of employee are involved, and the real or expected extent of their resistance.

This combination of forces for change (driving forces) and forces against change (restraining forces) are famously summarised in what is known as Lewin's equilibrium.

**Figure 4.2: Lewin's Equilibrium**

| Driving Forces (Forces for Change) | Restraining Forces (Forces against Change) |
|---|---|
| • Changing markets<br>• Shorter product life cycles<br>• Changing values towards work<br>• Internationalisation<br>• Global markets<br>• Social transformations<br>• Increased competition<br>• New technology<br>• New personnel | From individuals:<br>• Fear of failure<br>• Loss of status<br>• Inertia (habit)<br>• Fear of the unknown<br>• Loss of friends<br>From organisations:<br>• Strength of culture<br>• Rigidity of culture<br>• Sunk costs<br>• Lack of resources<br>• Contractual agreements<br>• Strongly-held beliefs and recipes for evaluating corporate activities |

Lewin called the process of balancing these driving and restraining forces a "quasi-stationary equilibrium", given that there is never a perfect balance between opposing forces, and thus always some movement taking place. The right conditions for change occur when this situation unfreezes, as occurs when restraints on the opposing forces are selectively removed. This will lead to change being pushed forward by the driving forces, until balance is once more achieved. The final stage is refreezing the new situation (from Wilson/Rosenfeld, 1990).

## 4.3 A Structure which Can Change Shape ...

It is said that the best way to ensure that a problem is forgotten about is to create a commission of enquiry. On this occasion, however, it is perhaps worth breaking with the tradition of government, and create a commission so as uncover the real issues. To demonstrate the benefits which this process offers, it is worth contrasting the planning and consultation processes.

The *planning process* consists first of all of drawing up plans, that is to say programming activities, which offer the greatest choice. The next step is to carry out at least some analysis of the future, usually referred to as *forecasting*. The final stage is to take decisions. Planning, therefore, is particularly appropriate for large businesses operating in a relatively stable environment.

The *consultation process* may have the same objectives, but achieves them with different and less heavy-handed methods, which are adapted to less predictable circumstances. The businesses for whom this book has been written are not large ones, and their environment is probably increasingly uncertain. The originality of the approach proposed here is that it includes a thorough examination of the general outlook, and of the opportunities and threats which it contains, and results in a strategic project which is built around the active strengths of the business, its long-term objectives, and its capacity to adapt.

This is why the Consultation Group should include the key decision-makers in the business, and not just technicians and those responsible for research and development. Equally, it should be open to anybody able to bring support, in whatever form. This being the object, it can both be highly structured (including a representative of each sub-group in the business) and highly flexible (enabling anybody who would like to participate to be able to do so, and making sure that this is well known).

It needs to meet at regular intervals, both during work time and at times which are split between work time and free time. Except in circumstances where adequate compensation can be offered, or the reasons for this are clear to all, the group should not meet exclusively in free time. The solution which asks participants to give up some of their own time, while including a significant granting of the company's time, is often the most effective. In the same way, it is useful if all the meetings can take place in a suitable (and the same) place, so that the project can be felt to be a specific activity, which creates and fosters a new attitude and way of working.

But this structure needs to be able to change shape. Discussion of the Diagnosis certainly requires the broadest possible participation. The gathering of documents, on the other hand, needs to be restricted to an appropriate group. The following chapters will describe the various stages in this process, each of which will require a specific and probably narrower form.

It's never an easy task to organise a group which is able to change shape. It requires a facilitator with a high level of tact and communication skills. There is always the risk that tension will develop between this group and the rest of the organisation: it may become dominant, or become marginalised, excluded or remote. It is necessary to ensure that anybody who is not asked to participate at a particular "technical" stage in the process doesn't feel excluded. Above all else, it is important to ensure that those involved in the process do not, progressively and almost naturally, take over everything else. It is here that communication of what is happening and following up what has been achieved by the project have a key role.

### 4.4 . . . While Nevertheless Following Precise Rules

To sum up what has been said above in a slightly different way,, any meeting which this consultation group holds costs the company money. So it is important to work fast. But at the same time it must remain as open and as flexible as possible.

All meetings of the group should be publicised to everybody in the organisation, all of whom should be free to participate and bring their point of view to the table.

Participation in the group is obviously linked to ability to make a significant contribution to it. It should also be anticipated that people will leave the group, so as to ensure that this does not take on unnecessary dramatic significance.

Each meeting should consist of asking general questions on each area of the Diagnosis, including those where a lack of information has left gaps.

This working group does not simply have the role of recording everything that it hears. It must have the authority to rework and adapt the Diagnosis and any of the conclusions which emerge from it.

How can the group best operate? It must, like any new structure, define at the outset:

- Objectives

- Length of its life

- Structure (facilitator, recorder etc.)

- Working methods.

It must also ensure that its role is well understood by the rest of the organisation; and it must be clear concerning its relationship with top management, and the impact which its conclusions can have. There is always some level of ambiguity in the role of such a group: so as to maximise its energy, it has to be seen as a leader; but, to be productive, it needs to have real freedom of movement and imagination, and thus needs to be able to take risks and ask searching questions of the business as a whole.

The key task of the Consultation Group, at this point, is to pursue a strategic thought-process based on the exchange of information and an interpretation of the Diagnosis. The operation of the group should lead, after several phases of activity, to the proposal of a number of options with a distinct technological edge.

The example below describes the running of a Consultation Group in three sessions each of three hours, as carried out in a business with 180 employees. It is included not as a blueprint, but as one way of going about the process which this chapter has described.

**Figure 4.3: The Three Meetings of a Consultation Group**

---

**Meeting 1: Understanding**

The starting point is the Diagnosis (as discussed in Chapter 3). The first meeting involves a pooling of information based upon a document put together by the assistant chief executive, which was previously circulated to the participants. The group is able to distinguish three particular types of problem which need to be resolved and to make a first evaluation of the forms of innovation (e.g. offensive, defensive) which could be used to meet the challenge.

This exchange of information enables discussion to take place and the Diagnosis to become more substantial. Out of a lively discussion there emerges agreement on a ranking of the strengths and weaknesses of the business and of its market position.

Following this, the group defines the possible ways forward which can be discussed at a second meeting. This consists of new questions which have emerged in specific areas. They have an impact on the employees in a variety of areas: borrowing capacity, market behaviour of certain competitors, possible productivity gains at the level of the organisation in the factory and of new approaches to the design process, etc.

**Meeting 2: Defining**

A fortnight later, a second meeting takes the form of a rigorous approach to strategy, which leads to an assessment of the range of possible future directions which the business can take. This includes the risks related to the behaviour of competitors and those who can reasonably be included in a broad definition of the business.

This begins with an analysis of changes and trends in markets and technologies, which is compared with an analysis of prospects for the business itself: financial base, investments, human resources and the organisation of production processes.

---

The objective is to target three possible solutions. These options are then listed and the likely necessary resources identified.

**Meeting 3: Proposing**
The object of this third meeting is to put the finishing touches to the document designed to aid in the decision-making process, and which will be presented to the management team. The participants will discuss each of the three options, criterion by criterion, but will resist any temptation to eliminate the one which appears to be the least satisfactory.

In this way, the group has been led to account for its proposals and justify its arguments by comparing the strategic objectives and the means of achieving them from one project to the others. Each of the possible strategic projects has thus been assessed against the following criteria:

1. Likelihood of local, national, international market impact

2. Technological feasibility (including how to get hold of the necessary technology)

3. Financial viability

4. Feasibility within the current state of the business (notably human resources).

At the conclusion of this particular exercise, the group was able to arrive at a double result which was communicated to the management team, and consisted of:

- A proposal made up of three precise and refined strategic objectives

- Calculations as to the resources which each of the projects would require.

This result was achieved due to the direct contribution of twelve employees (and indirectly of a good many more), as well as those of the company accountant, of an established customer and of a marketing expert in the sector (who was paid for this contribution). The group met again three weeks later, this time with the management team. It was on this occasion that a decision was made as to the strategy to be followed, which signalled the start of a new phase in the life of the business. Subsequently, the group met from time to time, and was to be reactivated two years later to look at how a new product launch should be undertaken.

# SELECTING A STRATEGY

*A good deal of space has been given to describing how to prepare and carry out a Technological Diagnosis for the business, so as to be able to assess and compare the risks associated with a number of different options. A flexible approach to the consultation process enabled the creation of a larger group with the responsibility for coming up with possible strategies, and testing them out before they are put into operation, in an exercise involving the maximum number of employees. Beside the usual outcomes, the process always results in a number of unexpected discoveries.*

*Both the ranking of the strengths and weaknesses of the business and the identification of visible opportunities may have an immediate impact in management terms, but they cannot be seen as a substitute for setting the strategic direction of the business. It is now time to turn to the decision-making process, which is very much central to the activity that this book is describing. This is all the more so since these decisions cannot normally be reversed, involving as they do major changes such as a significant innovation. This chapter looks at this type of decision.*

*A strategy is more than just an automatic extension of recent trends: it is a structured thought process which involves following a limited number of clear guiding principles over a probably lengthy period. We can talk of strategy, for example, where a business sets itself the objective of launching a new product onto the market every five years; or when a decision is taken to diversify into two distinct areas of activity; or to launch a new vaccine; or to look into a new market overseas; or to change the nature of the business. But it is also a question of strategy when considering the image that one wants to give to the company, or when a decision is taken radically to change the nature of working relationships within the business. In the majority of cases, this decision is, in the last analysis, that of the managing director. In a few particular situations, however, this is a joint decision; and the recent trend is for consensus increasingly to be the case. But the process remains the same: it is necessary, in the end, to take the plunge, and to take the decision which initiates a major innovative leap.*

*The different stages in the process of arriving at a decision necessarily include consultation, calculation and analysis. Haste is never justified if all that it produces is an unsound decision. In particular, the kind of decision to which this chapter refers does not involve the once-and-for-all establishment of a detailed project plan. What matters here is the strategic decision itself: whether or not to go for a new product, new piece of machinery or new working system. The decision is strategic because it commits the organisation to a particular course from which it cannot turn back.*

*It represents therefore a choice of direction, and requires motivating the key energies within a business to pursue a precise and ambitious objective, yet knowing that the other areas of business activity must not be neglected; and knowing as well that the ways in which implementation will take place have yet to be established or have every chance of being substantially modified several times in the course of the pursuit.*

## Figure 5.1: Selecting a Strategy

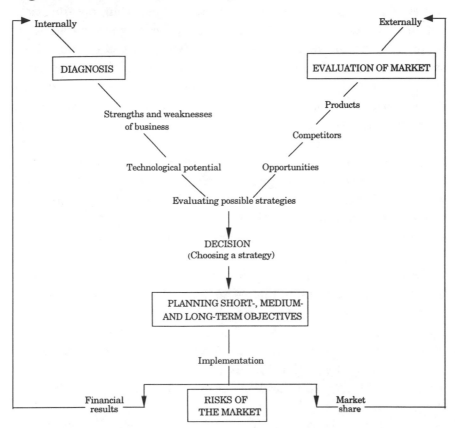

### 5.1 Consulting and Asking Questions

Any decision-making process is both progressive and iterative: it has to be gone through several times before you arrive at a result. It begins with an analysis of everyday reality (simplified to concentrate on the essential), and measures and compares a number of different opportunities, deliberately raising as many opposing views as possible so as to be certain not to ignore a key aspect of the issue under examination. At the same time, therefore, it is both critical and constructive.

The next stage involves scanning the different factors which make up the future: expected competitor behaviour, advances in science and technology, substantial changes in patterns of demand, exceptional ideas which the strategists or their colleagues have just come up with. This is all very much an area of uncertainty. Carrying out a good evaluation of different possibilities, which examines their likelihood and the costs associated with them, is certainly one of the most difficult stages of the whole strategic process. The next stage is the overall decision, the decisive strategic choice. It needs to be seen as *a single point in time, amid the complexity of daily activity*. For an instant, everything needs to be seen in terms of "all or nothing" or "it will be this way rather than that way". It is obvious that such a decision needs to be carefully considered in the context of the reflection and analysis that has gone before, all of which should be used as a tool to assist with the making of a decision. But the final decision has to be taken relatively alone, in the realisation that it will have a key impact on the future success or failure of the business.

This stage of the process is a watershed, because it is very difficult — and often impossible — to undo the decision once it has been made, communicated, and the resources committed for its implementation. In the case of failure, the business could well go under, or be forced to make another fundamental (and probably defensive) transformation, involving at the very least the loss of all the credibility, time and investment (sunk costs) devoted to the project.

So once the strategic decision has been made, the problems don't simply vanish. Each of the subsequent questions which needs to be resolved may seem to be less critical, but there are many more of them. From this point onwards, the process inevitably becomes more complicated, since implementation requires a strategic decision being cascaded throughout an organisation which has been established for some time and where everybody has their own way of doing things. Everyone in the business, whatever their area of responsibility, is going to be involved in taking important decisions in their area of activity, and all of these smaller decisions need to fit in with the central strategic decision.

The first stage in making a decision is asking the right questions. An earlier passage described how making a strategic choice needs to be the culmination of a maturing process, which should consist of a technical study (the Diagnosis) and a consultation involving a number of people. These different stages involve an effort to understand individual experiences and opinions, to define policies and to compare different possible solutions. This needs to take place across the areas of technology, of finance, of production, of the market and of employee behaviour.

## Figure 5.2: Some Preliminary Questions

**What is a major innovation actually going to do?**

- Resolve a major problem facing the business: *defensive innovation*
- Ensure growth in years to come and enable the business to get ahead of its competitors (first move strategy): *offensive innovation*
- Enable the business to keep up with its competitors: *imitative innovation*
- Respond to an opportunity which won't arise again: *opportunity innovation*
- Exploit a scientific or technological advantage in use elsewhere, by acquiring patent rights (for example): *technology watch innovation*

**Where can an innovation have an impact?**
- On the design and production of products
- On equipment
- On new materials to be used
- On the knowledge base of the business
- On the organisation of production
- On the organisation of administration
- On the organisation of suppliers
- On the whole organisation

**What methods can be used?**
- Exploitation of knowledge within the organisation
- Acquisition of patent rights
- Acquisition of external expertise
- Development and implementation of a training plan
- Purchase of materials or machines which incorporate new technologies
- Recruitment of someone with particular technical knowledge

**How can real change be brought about?**
- With the available technology
- With the shape of the business
- With effective customer demand
- With the commitment of employees and other partners in the business

These questions, which are established at the very start of the consultation process, can obviously be refined and added to. The important point is that *they should ensure that everybody is convinced in each case with regard to the evidence, to likely changes in the future, and to the opportunities and risks*. There will always be areas of uncertainty and ambiguity, given the limitations of a narrow "national" way of thinking as well as the impossibility of mastering all areas of knowledge and information. But the more thoroughly the preliminary work has been carried out, the more likely it is that the answers to these questions will be more than simply black or white.

### 5.2 Measuring, Interpreting, Eliminating

The above list of questions forms the first stage of the strategic thought process, and provides the necessary basis upon which the criteria for a final decision can be established. It is not, however, simply a question of creating a "recipe for decision-making", in which, provided the right ingredients are included, the correct answer will always emerge at the end of the process. So as to illustrate the ways in which different criteria can overlap, the pages which follow present several classic "partial models", each of which can be used to throw light upon perhaps one area of the questions which need eventually to be answered. These models are those of the product (and business) life cycle; of the technology life cycle; of the Boston Consulting Group model (which is usually known as the Boston Box); and of the 2 x 2 matrix. Each of these models has known its moment of glory; and has then been criticised (usually in direct proportion to its excessive and mechanical use by consultants), only to be replaced by even more powerful and complicated models. This is certainly the case of the celebrated models of Michael Porter, who has taken up a number of ideas in this area developed at Harvard and in major European business schools. But this chapter will concentrate at this point on the classic models, on account not only of their simplicity but also because they allow the user to design a working synthesis which draws on them all.

### *The Business Life Cycle Model*

This is a long-established concept, first advanced in 1890 by Alfred Marshall, a British economist. It goes like this: businesses develop and encounter problems which can often be compared to the human life cycle of birth, growth, maturity, ageing, death. The key difference is, however, a substantial one: for a business, death is not inevitable. At certain points, it is possible radically to change the shape of the business, and to alter the way in which it operates, so as to restore its youthful vitality.

### Figure 5.3: The Business Life Cycle Model

---

**Phase 1: The business as project**
- The business does not exist yet
- Constructing plans
- Nobody has much experience
- Very little money is needed (personal or family money, or sunk funds)
- Much advice is needed

---

**Phase 2: Creation or launch**
- The project moves from being an objective on paper to becoming a first attempt at definition (prototype)
- The innovation may be in high technology, but it remains relatively simple
- The business needs more money, with a very high level of risk
- It requires a lot of determination, sense of proportion, team spirit . . . and a substantial chunk of luck

**Phase 3: The development gap**
- This is the critical period in the life of the business
- The project is growing fast
- The need for finance is growing faster than the generation of income
- There is always a gap between the hope of success and the success itself, due to unexpected delays and sales which don't materialise
- There are always more unexpected problems than unannounced windfalls
- The whole project is in danger of being aborted (this is the phase with the highest business death rate)

**Phase 4: Growth**
- Profitability has arrived at last
- The market has grown
- Further investment is urgently needed to meet increased demand
- The business borrows and can repay its borrowings
- The more the business invests, the more it is possible to benefit from economies of scale and the learning curve
- Success breeds success

---

**Phase 5: Managing maturity and restructuring**
- The market is getting saturated
- The business understands its sector, but it is finding it increasingly difficult to challenge its own way of doing things
- More cash is being generated than is required to grow in its own sector

**At this point several options appear:**
- Become a cash cow for other activities of the business (see the Boston Box, below)
- Try some radical innovations in the processes used
- Sell the business before it becomes apparent that it is past its peak
- Totally reshape the activity of the business.

---

### The Technology Life Cycle

The same cycle which businesses experience can equally be applied to products and technologies. Lip, the French luxury watch producer, was caught out by the appearance on the market of quartz watches, and went into liquidation. But a few years later Swatch managed to launch a new generation of watches and so create itself a market share, in the face of aggressive Asian competition. In other sectors, there are no longer any producers of slide rules, of valve radios, or of wind-up gramophones to be found. Substantial fortunes were made, however, out of these products. These changes, in terms of rise and fall, are the direct consequences of major technological movements.

It is thus possible to apply the life cycle model to many aspects of human organisations, and to technological projects in particular. A number of economists have done so in the wake of work carried out at Harvard by Hayes and Wheelwright. They argue that according to the degree of maturity of the production processes, one can produce in small runs, in long runs (mass production), or using continuous process technology which combines the maximum standardisation of components with the complexity and the increasing differentiation of the end product.

This argument is illuminating, but it should be used as a generator of ideas, rather than to explain everything. Beyond this point, there remains considerable margin for free expression, and no shortage of cases which contradict this fairly rigid model. One of the improvements which has been proposed involves the division of the technologies into three distinct categories:

- **Standard technologies**. These are generally to be found in all businesses. Different businesses may reach and use them at different stages, but they are not the means to achieving differentiation from the competition. This is true of word processing, of clutch and shock absorber systems in cars, and of transistors or cathode tubes.

- **Key technologies** are those which today determine the differences and the competitive advantages which one business possesses with regard to its competitors. This could be due to computer software, to new materials, to the systems which it has introduced to save energy, or to digital imaging. These technologies will probably only play this role for a short period. They will then either be abandoned due to their failure (e.g. the tilting train) or they will become standard because everybody in the market has adopted them (e.g. the fax machine).

- **Future technologies** consist of an enormous number of concepts which substantially alter the balance of power between companies in favour of those which harness them, as long as these technologies turn out to be appropriate. This is true for voice recognition, for high-definition television, for new materials for use in high temperatures or for superconductors.

### The Boston Consulting Group (BCG) Model

Invented for multinationals, this model was flavour of the decade from the mid-1970s to the mid-1980s. It creates a theory of industry strategy which is based upon the balance between two marketing indicators: the market share which the business enjoys compared with that of its competitors, and growth in the total market. It enables one to situate different business activities and to compare them with each other while identifying potential synergies, essentially in terms of productive capacity and of financial demands. It is healthy for a business to carry out a diverse range of activities which achieve a balance between those which generate capital and those which demand it. This matrix also highlights the best reaction when faced with newly emerging products, or with those which are coming to the end of their life.

**Figure 5.4: The BCG Model**

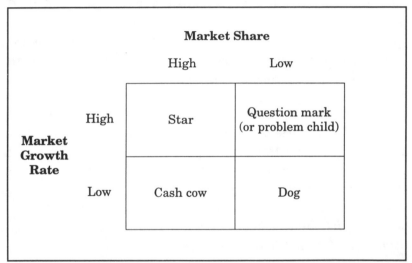

- **Dogs** have a low share in static markets and can thus be considered to be the worst of all possible combinations. They are often a cash drain and may use up a disproportionate amount of company time and resources. However, a dog may have what is in fact a low share of a future market, thus providing a real opportunity for innovation, although this may well be a high-risk strategy.

- The **question mark** (or problem child) is in a growing market but does not have a high market share. Its parent company may be spending heavily to increase market share but, if this is the case, it is unlikely that sufficient cost reductions are being achieved to offset such investment. This is because the experience gained is less than for a star and costs will be reducing less quickly.

- A **star** is a product (or business) which has a high market share in a growing market. As such the company may be spending heavily to gain that share, but the experience-curve effect will mean that costs are reducing over time and hopefully at a faster rate than the competition. The product could then be self-financing.

- The **cash cow** is a product (or business) with high market share in a mature market. Because growth is low and market conditions more stable the need for heavy marketing investment is less. But high market share means that experience in relation to low share competition continues to grow and relative costs reduce. The cash cow is thus a cash provider.

The main problem associated with the Boston Box is that it does not reveal the coherence which may exist between the different activities within the business. Nor can it take into account the synergies between related industries, nor the extent of market and technological fit. Through concentrating too much on financial return, market strategy is somewhat overlooked. In short, *what is right for the highly diversified multinational may not be right for the smaller business*.

### The 2 x 2 Matrix
Ansoff's strategic portfolio matrix describes four distinctive strategic business areas, each of which offers different opportunities for growth and profitability while demanding different competitive approaches, as shown in Figure 5.5.

- **Market penetration** represents growth through increased market share for the present products, by making (for example) minor improvements. This is not an obvious source for innovation opportunities, although it is relatively low risk.

- **Market development** involves greater risk than market penetration, and often leads to failure where the different needs of the new market are not sufficiently recognised.

- **Product development** creates new products to replace current ones, and by aiming at established customers can build upon a firm's knowledge of its market. This area includes internal process innovations. Risk in this area is low-medium.

**Figure 5.5: The 2 x 2 Matrix**

| Mission \ Product | Present | New |
|---|---|---|
| Present | Market penetration | Product development |
| New | Market development | Diversification |

- **Diversification** is distinctive in that both products and markets are new to the firm. This is very much the high risk (and therefore potentially high return) option.

Research carried out by Cranfield in 1992 into the marketing strategies of small companies identified 45 per cent opting for market penetration, 25 per cent for market development, 25 per cent for product development and 5 per cent for diversification.

The tools described below, although they have all been in and out of fashion, nevertheless remain a valuable source of ideas and discussion. They certainly provide a justification for chosen strategies and they enable one to develop a workable model of their consequences.

**Figure 5.6: Strategic Options for Smaller Companies**

| Development Directions | Development Methods | | |
|---|---|---|---|
| | *Internal* | *Partnership* | *Acquisition* |
| **1. Do nothing** | | | |
| **2. Withdraw** | • Liquidate | • Licensing <br> • Subcontracting | • Complete sell-out <br> • Partial divestment <br> • Management buy-out |
| **3. Consolidation** | • Grow with market <br> • Increase: quality, productivity, marketing <br> • Capacity reduction/ rationing | • Technology transfer <br> • Subcontracting | • Buy and shut down |
| **4. Market penetration** | • Increase: quality, productivity, marketing | • Collaboration | • Buy market share <br> • Industry ra- tionalisation |
| **5. Product/ service development** | • R&D <br> • Modifications <br> • Extensions | • Licensing <br> • Franchising <br> • Consortia <br> • Lease facilities | • Buy in products |
| **6. Market development** | • Extend sales area <br> • Export <br> • New segments /new uses | • New agents <br> • Licensing <br> • Consortia | • Buy competitors |

But more recent work into the life cycle — and thus the maturing process — of technologies has underlined several new models, and in particular the three which are described below.

1. **The changing importance of product innovations and of process or organisational innovations**. When a new technology appears, attention is focused on its physical capabilities and on the improvements they bring, whether in the case of the jet engine or of the printed circuit. Business

growth and success are at first related to the new applications which can be found for the new product. But, as the technology gradually moves to the point of successful application and mass production, then the nature of the problems changes. At this point there is a much greater interest in the different components which make up the product, and in the details of its manufacture. Improvements become more difficult to achieve and more limited in their impact. So *productivity gains in process terms become the prime objective*.

So the birth of an innovation leads to a product achieving dominance through its technical features; growth necessitates improvements in the processes of its manufacture, and maturity sees a concentration on the commercial and financial dimensions of its production. But in any case, *there is no technological innovation if there is not an innovative organisation*.

2. **Technology improvement**. The life cycle of a technology is never entirely determined in advance. The introduction of incremental improvements can substantially extend an innovation's life and enable it to retain its advantage over its competitors despite the appearance of imitators. Within any industry, it is the leading company which is first able to gain access to new information and to improve existing processes. So, in many cases, a given process or a given product can be revitalised a number of times by keeping the core unchanged and by innovating with regard to another of its components.

3. **Major technologies and the lack of continuity in the rhythms of innovation**. Major innovations appear one at a time rather than in a constant stream. The technological leaps forward which they cause raise the question of the choice of substitutes and the strategy to adopt when faced with two technological competitors. To what level should one attempt to improve an existing technology? What investments should be made to gain control of the new technology? When should one replace one by the other? So one of the key strategic issues is the balance to be struck between the pursuit of perfection in the traditional technology and the search for substitute tech-

nologies. While the sailing boat resisted the steamboat for fifty years, it only took the electronic calculator five years to kill off the slide-rule. So the choice that needs to be made by each business looking to adopt a major innovation from outside (a new piece of machinery, for example) should be based upon a comparison of the unit production costs of each technology.

## 5.3 Deciding on a Strategy

It is obviously important to have the resources to implement a strategy, but first of all it helps to have a strategy. The two are very closely linked. Up to this point, much space has been devoted to the resources, and to outlining some of the theories upon which a choice can be based. But the time has come to take the decision which will define the policy guidelines to be followed in the future.

There is a fundamental difference between the preceding stages and this one: it lies essentially in the contrast between the consultative nature of all that has gone before, and the solitude of one person making a choice, even if it is only for an instant. *The more the decision is strategic, unique and irreversible, the greater the sense of isolation of the decision-maker*. Nor is this slightly military description of the decision very far from the truth. This is not only because it is often the decision-maker who will finance the resulting project; but is also due to the image of the decision-maker being forced to choose between several possible options — including one which is to do nothing. And however much preparatory work has been carried out, the decision-maker invariably suffers from limited or inadequate information. The decision concentrates into an instant the substantial process which has for the most part taken place elsewhere, and has involved a large number of partners (some of whom have probably been unaware that they have participated in the process at all).

*The decision itself is thus the key point of the whole process*. No one pretends that, in the majority of cases, it will be easy or even possible to retreat from this decision or to change direction in mid-stream. If this does become necessary, it will be sure to bring with it the burden of substantial *irrecoverable costs*: all the time which has been devoted to preparing the project, the expense of the initial study, the investment already made in this particular project, the organisation and reorganisation which has already taken place. And beyond this lie all the expectations which have been growing in the minds of internal and external partners, who will inevitably be profoundly disappointed, and whose confidence in future projects is likely to be seriously undermined. The costs of all these is always far greater than their actual financial cost.

There is no formula which enables the right decision to be made every time. The only compensation for the responsibility of the decision-maker ought to be the right to make a mistake. This exists in theory, but where strategic decisions are concerned it is impossible to escape from the extent and duration of the impact of the consequences of a wrong decision. So it is important to know how to put off making a decision for as long as possible, until the decision and the options are in sufficiently sharp focus, and until there has been thorough consultation with all partners who might have a different view on the subject.

In the last analysis, the quality of any decision depends on a long list of factors and imponderables over which one has no real control (such as luck, and the intuition of the decision maker). But these can be shaped, or at the very least counter-balanced, by those factors over which it is possible to have an influence: the incisiveness and quality of the Diagnosis, involving a detailed analysis of the market and its opportunities, and allied to a detailed calculation of the costs.

*A good strategy can create success out of initial weakness*. There are four crucial elements that all successful strategies share:

- They are directed towards unambiguous long-term goals

- They are based on detailed knowledge and understanding of the external environment

- They are based on an honest assessment of the organisation and its capabilities

They are implemented with energy, perseverance and commitment by everyone in the organisation.

# Chapter 6

# PREPARING THE ORGANISATION

Contents

6.1   Innovation and the Organisation

6.2   Innovation and Information

6.3   Innovation and Employee Involvement

6.4   Innovation and Training

*The aim of the last three chapters has been to prepare the ground for a decision on the form and goal of an investment in a major innovation. At the same time, it has described the process through which any innovation and the related technologies can be accepted by every part of the organisation. This chapter looks at the inevitable impact of such changes on the organisation, on its information systems, and on employee involvement and internal training. In every case, the issue is one of people management. This discussion is a necessary preliminary to the actual implementation of the innovation, which will be covered in Chapter 7.*

*The two key elements in any major strategic decision are organisation and technology. The link between them highlights the fact that a technological system is well designed where it uses and maximises both the potential of the organisation as it is now (its flexibility, quality, productivity) and the capacity of its human resources (skills, knowledge base, creativity, problem-identifying and -solving ability). Furthermore, an organisational system is well designed when it does not create obstacles to the use of available technologies, when it makes use of the most highly-*

*developed areas of the business, and where it leaves a space in which creativity can flourish and change mature.*

*It is important, however, not to lose sight of the big picture in the struggle to resolve points of detail. Any plan to improve the productivity or efficiency of particular jobs (to reduce machine time, for instance) needs to fit in with the benefits in quality and flexibility which apply across the whole organisation. In the same way, it is necessary to work to extract the maximum benefit from every opportunity for technological integration. For example, a computer-aided design system (CAD) must not only be seen as a way of replacing the drawing board and thus reducing design time. It can also be the first link in a more ambitious computer-aided manufacturing system (CAM).*

*It will next be necessary to identify the changes which need to be made to manufacturing processes, to the ways in which work is organised, and to people and their roles. In particular, it is crucial to consider training needs, and how they will be met and evaluated. There are two particular reasons for this:*

- *The first, obviously, is to ensure that the introduction of change is related to improved effectiveness, so that the "hard" part of innovation (the technology, the machine) fits the "soft" part (the human resource base);*

- *The second takes a much wider definition of effectiveness, and involves ensuring that the whole of the organisation accepts the innovation process. With this objective, the innovation process needs to ensure maximum coherence between the interests of the company and those of its members.*

*For these reasons, training needs to be seen as part of a broader strategy, and thus as indispensable to the process of leading everyone in the organisation through change. It can also be seen as an aspect of the non-financial rewards of the process as a whole (reduction in direct labour, work becoming more intensive, changes to specific tasks, moving from job to job).*

*In other words, training must not be seen simply as an increase in overheads, as is too often the case, but as an efficient and cost-effective tool to increase the commitment and skills of every employee. Any increase in the total knowledge base has the double*

*benefit of adding value both to the individual and to the organisa-*
*tion as a whole. There is always the risk that the trained employee*
*will leave the business to take a better paid job elsewhere, either of*
*her own volition or through being headhunted. This is a very*
*common practice, particularly in areas or industries with less de-*
*veloped training networks and skill levels. But the business can*
*reduce this wastage by making sure that it remains more*
*"attractive" than its competitors in the labour market. And an*
*ideal way to increase its own appeal is to involve all its employees*
*in its innovation strategies.*

## 6.1 Innovation and the Organisation

Traditionally, an organisation consists of formal networks which enable the different functions in the business to express themselves (e.g. managing director, finance, human resources, supplies, technology, marketing, etc.), with the goal of these networks being to minimise costs and to improve efficiency, effectiveness and market share. The organisation is thus structured around the idea of reason, of order, and of the performance of a series of related activities in the framework of a division of labour. This is reflected in the hierarchy, in which it is effectively the top of the organisational structure which decides how the business should be organised and how it should be developed; and also the place of each individual on the organisational chart which maps the network.

Today, these functions have substantially changed. The organisation must respond to new external forces which have been whipped up by the rapid change that is taking place in the environment. More than this, the organisation must itself become involved in these changes, and create its own response mechanism so as to participate within the most influential of them: instability, permanent change and thus uncertainty. A business can no longer restrict itself simply to continuing to do exactly what it has done in the past. It must itself become a source of change, of new initiatives, of innovation.

But the way in which the organisation relates to and handles change is not a rigid one. There are two ways to react. The first involves planning for change, through R&D action programmes, and the acquisition of technology. While the second option consists of creating the space in which change can occur, while leaving it to the individuals to take the initiative: creative space; the freedom to organise one's own work, with perhaps 10 per cent of one's time free to be used in the way that the individual considers to be most valuable. Where major innovations are required, the planning-for-change route is certainly the preferred one; but where a number of small innovations are the desired outcome, the space-for-change option is more likely to bear fruit. In all cases, however, it is necessary to envisage substantial change in the following areas:

- The shape of organisational chart (the Consultation Group is an example of this);

- The tasks which each employee carries out (with an increase in versatility, and clear implications for the specialists);

- Relationships involving collaboration (and therefore also conflict).

This approach to organisation certainly implies complexity rather than uniformity. Its objective is to minimise the "transaction costs", that is to say the time and energy which need to be expended for every individual to keep in touch with what everyone else is involved in; and to increase loyalty and effectiveness in working relationships with both internal and external partners. For any major innovation creates disruption in the existing order of things. So any decision intended to change existing working relationships needs to act on all of these variables at the same time, and the introduction of a new technology involves a re-examination, albeit it in some cases partial, of every component of the internal and external networks.

It is, however, possible to ensure a fit between the different variables which have an impact on the organisation and the nature of innovations and the corresponding technologies. A flexible technology, for example, is easily assimilated within a flexible type of organisation or network, and within a co-operative style of human resource management; whereas a rigid technology is more usually associated with hierarchical organisations and with a more authoritarian style of people management.

An analysis of two typical but opposing types of organisation illustrates this argument: the operating organisation and the innovating organisation. The figure below presents the characteristics of each organisational model. From these it is possible to estimate the assumptions and the constraints which exist within the structure of the business, within the structure of work and within the environment, all of which influence the choice of the most appropriate actions required to guide the innovation process in a specific business.

One of these models is drawn from the rational, bureaucratic structure which one might encounter in the engineering sector; while the other is perhaps a typical information technology culture. In the real world, of course, the nature of a business will never be this clear-cut. It will most likely be somewhere between the two as a result of the impact and constraints imposed by the environment, structure and technology. But, at the very least, close examination of the business and its situation can identify the room for manoeuvre which it enjoys.

**Figure 6.1: The Maintenance and Innovation Models: Two Types of Effective Organisation**

|  | Maintenance | Innovation |
|---|---|---|
| Culture | • A role culture, governed by precedent<br>• Routine operation, low uncertainty<br>• Stable | • A task culture<br>• Problem setting, high uncertainty, high turbulence |
| Structure | • Bureaucratic, division of labour, hierarchic | • Flat organisation<br>• Task-oriented project teams |
| Processes | • Operating units controlled by top management with a strategic/operational/ financial planning role, measuring performance | • Process directed towards generating, selecting, blending of ideas<br>• Flexible strategic planning, operation controls loose |

|                     | Maintenance                                                                              | Innovation                                                                                           |
| ------------------- | ---------------------------------------------------------------------------------------- | ---------------------------------------------------------------------------------------------------- |
| Rewards             | • Pay and benefits. Promotion up the hierarchy<br>• Status symbols                       | • Learning, recognition.<br>• Equity participation in new ventures<br>• Bonus payments (reward by results) |
| Leadership          | • Hierarchical and established in advance                                                | • Drawn from any level according to each case                                                        |
| Relationships       | • Usually top-down                                                                       | • Also bottom-up<br>• Plus networking<br>• Co-operative                                              |
| Strategy            | • Medium-term, centred on production                                                     | • Long-term, centred on the market                                                                   |
| Financial Criteria  | • Economies of scale<br>• Lower transaction costs                                        | • Economies of flexibility                                                                           |
| People              | • Recruitment based upon needs of the structure for specific skills                      | • Need for idea generators who combine technical knowledge with creativity                           |
| Communication       | • Constrained, vertical                                                                  | • Open, lateral                                                                                      |

Every major innovation is going to change the organisation. But to what extent is the business capable of making the change required? Any change needs to take into account the constraints on the organisation as it is now. So how far can the business go? The table below shows that it is possible to calculate the degree of change that is likely to be involved.

**Figure 6.2: Levels of Organisational Change, Classified by Degree of Change**

| Degree of Change        | Operational/ Strategic Level                      | Characteristics                                                                           |
| ----------------------- | ------------------------------------------------- | ----------------------------------------------------------------------------------------- |
| Status quo              | 1.  Can be both operational and strategic         | No change in current practice                                                             |
| Expanded reproduction   | 2.  Mainly operational                            | Change involves producing more of the same (goods or services) and incremental innovation |

| Degree of Change | Operational/ Strategic Level | Characteristics |
|---|---|---|
| Evolutionary transition | 3. Mainly strategic | Change occurs within existing parameters of the organisation (e.g. change, but retain existing structure, technology, etc.) |
| Revolutionary transformation | 4. Predominantly strategic | Change involves shifting or redefining existing parameters; structure, technology and organisation of labour likely to change, for example |

## 6.2 Innovation and Information

The previous section examined the different organisational models in a fairly abstract way. It is now necessary to consider the consequences for the information systems of introducing a major innovation.

It should by now be clear that any new approach leads to organisational changes. And any technological change, even if it is limited to the physical layout of equipment (and particularly if it involves changing the hardware or the software of an existing technological system) must take into account the way in which the business is currently organised, and move forward keeping this clearly in mind. There are numerous revealing case studies, on the other hand, which describe workshop reorganisations which failed because the equipment had been moved around without the workforce being consulted.

Consider the case, for example, of the introduction of a flexible manufacturing system (CAD/CAM) into an assembly line workshop: the change just didn't take root. Such a system requires a certain quality of work, of capacity to adapt, of communication with the other operators, and of creativity which the Taylorian system is unable to generate. If it is to succeed, a project of this kind needs to develop, in advance, a training structure, and the variety and enrichment of work for the operators involved. But it must be accepted that this may involve a change in the culture of the organisation.

There is the counter example, however, of a more organic structure, such as a flexible and fast-reacting small company, which needs to develop a network of subcontractors so as to increase production when faced with unexpected and sudden market demand. This change alters the logic upon which internal working relationships are based, and demands careful handling, notably with the internal workers. One solution in this example could be to create a network of small producers, enabling those who had previously worked inside the organisation to create their own small business.

The organisation of innovation is thus a continuously evolving process, which demands control of an increasingly complex situation. It involves developing a learning culture to the point at which it becomes a strategic strength of the business.

To illustrate the consequences of these choices in action, it is useful to take the example of the introduction of new technologies in the area of information.

**Figure 6.3: Organisational Change and Information Technologies**

| Technology | Type of organisation |
|---|---|
| 1. **Architecture of the structure of the information system** | **Control and co-ordination system** |
| a) Centralised | By procedures and plan |
| b) Spread | By results |
| 2. **Use of software** | **Structure of tasks** |
| a) Totally automated | Human tasks secondary; passive monitoring |
| b) Dependent on central control | Automatic control system |
| a) Centred and rigid | Human tasks in sequence |
| b) Off-centre and interactive | Continuous tasks |
| a) Neutral interfaces | Specialist tasks |
| b) Co-operative interfaces | Discretional tasks |
| 3. **Communication system** | **Decision-making process** |
| a) Top-down or bottom-up | Hierarchical |
| b) Interactive | Co-operative |

This example illustrates the truth that for the introduction of one or more innovations into a business to be effective (and not merely efficient) a process is required which does not end with the

Diagnosis and the taking of a strategic decision. It very much continues throughout the pilot phase of the innovation project itself. This phase always turns out to be more delicate than was anticipated; so it is better to overplan it rather than underestimate it, and to invest in its success the most highly qualified resources, both internal (management team and Consultation Group) and external (experts, agents for diffusion and technology transfer, service centres).

## 6.3 Innovation and Employee Involvement

The impact of innovation on the culture of an organisation is considerable, and needs to be planned for and evaluated across the areas of social planning, employee involvement and the training of all staff (including the top managers).

---

### *Antoine Riboud (former President of the multinational BSN) Defines Social Planning*

*"Social planning is a vital necessity which has tended to be undervalued in France. A number of large companies already carry out a form of labour market forward planning, although this is mainly a statistical exercise which looks at the age spread of the workforce, at the average age, at the flow of people out of the organisation, and the qualifications of those recruited.*

*This is certainly a necessary first stage to any form of prediction, a kind of demographic snapshot, which ensures that the business has sufficient human resources to be able to meet its targets.*

*But it is important to go a good deal further, into areas which most companies rarely enter, and consider the possible change which could occur in a number of areas: in skills, in the way that work is organised, in training needs and practices, in work stations. All of this needs to be examined in the context of anticipated technological change, and changes in the nature of demand as established market segments develop, merge or vanish altogether.*

*This kind of qualitative social planning is anything but a naive exercise in fantasy. Its object is to involve all managers in the process of planning for the future. It is in this way that the organisation can respond to the demands of the production director who prefers automated mass production, using a small, multi-skilled workforce; to the demands of the marketing director who insists that highly-segmented demand requires flexible production lines, which involves a larger workforce who are able to experiment rapidly and produce new products; to the demands of the finance director who wants to reduce new investment and maximise the return on what is currently in place; to the demands of the personnel director who wants to keep control of the career opportunities open to the staff involved in production; and to the demands of the IT director who is keen to rationalise the use of data processing and office automation (Riboud, 1987).*

---

The participation of the workforce in the introduction of innovation needs to take account of:

- Training

- Industrial relations

- Job security.

The focus of this book means that the training question will be given particular attention. But the other aspects are equally critical to the success of any strategy designed to ensure maximum commitment. There are numerous examples of projects which have failed precisely because training plans were constructed and implemented without taking industrial relations into account, or without paying sufficient attention to the existence or otherwise of ways to keep people happy. There are certainly a number of potential forms of participation available, ranging from the information which is made available to individuals, via production conferences and quality circles, to innovation contracts involving profit-sharing.

Ideally, a range of methods should be used — they are certainly mutually reinforcing rather than exclusive. But involving the workforce is a strategic action, and neither its complexity nor the amount of time or planning it demands should be underestimated.

In the introduction of any substantial innovation, there are a number of key points at which the participation of the workforce is essential:

- During the stage at which the technologies are selected

- During the stage at which the innovation is introduced

- At the point at which it is necessary to develop technical solutions and to adopt the initial organisational shape.

The development of the strategy to ensure the participation of the workforce can best be allocated to the Consultation Group which was set up in the earlier stages of the innovation project (see Chapter 4). This may well involve moving from the limited participation of a small project group (composed of technicians and "leading" employees) to the participation of the workforce as a whole, including those from elsewhere in the organisation who are the most likely to be the first to experience the impact of change. This involvement is crucial, and the earlier this group is brought into the consultation process, the more likely the eventual success of the whole project.

### *Information Made Available to Individuals*
Nowadays, a growing number of medium-sized enterprises involve their employees in general, and their technicians and managers in particular, in training activities. For example, of a representative sample of 460 small-medium enterprises in Lombardy who were analysed in a wide-ranging enquiry during the preparation of this book, it emerges that manufacturing companies with fewer than 100 employees provide training for their machine operators (30 per cent of the cases) almost as often as they do for their managers (39 per cent of the cases). The training provided for the machine operators is mainly in the area of the introduction of innovations into office life (IT skills) and the use of the machines used in production (numerically-controlled). Such activity is found particularly in businesses using current technology in which the majority of the workforce are not directly involved in innovation (standard production using non-qualified labour). Those who

benefit are the machine operators, since the rest of the organisation does not really have to adapt its behaviour.

Management training is another area of individual training. This is particularly to be found in large companies but is now beginning to extend to the smaller companies. In the majority of cases, it is necessary to use business schools, or distance learning programmes, to make headway, although a recent study in West London identified a growing number of small firms making use of Business Link management training programmes, or preferring to use consultants as management developers rather than on specific problem-solving projects. This study identified particular weaknesses in management in the following areas:

1. Strategy formulation

2. Performance measuring and reporting

3. Human resource management

4. Planning and budgeting

5. Insufficient knowledge of key technologies used

6. Listening to suppliers, customers and competitors.

These are not the only activities of managers (and need to be read in conjunction with those in the box below). But they are all, even the last of them, highly-structured activities that are important elements in the change process; and there is little sign in the majority of small firms that top managers accept the extent and urgency of their own training needs.

**Figure 6.4: Key Management Tasks**

- Fast reaction
- Estimating capacity
- Estimating costs and delays
- Breaking bottlenecks
- Systematising diverse elements
- Developing standards and improving methods
- Balancing process stages
- Managing large, specialised and complex operations

- Meeting material requirements
- Running equipment at peak efficiency
- Timing expansion and technical change
- Raising capital required.

### *Production Conferences, Quality Circles and Innovation Contracts*

Training is not limited to the areas of sector-specific or functional skills. The pursuit of quality has led to the creation of quality circles, seminars, and other activities with total quality management (see the box below) as the objective.

### Figure 6.5: Total Quality Management (TQM)

TQM is not simply a question of employing more quality inspectors. It involves management commitment, a quality assurance system, quality tools and techniques, and teamwork. TQM can best be achieved, it has been argued, through the following:

**Effective leadership**
- Develop clear beliefs and objectives
- Develop clear and effective strategies
- Identify critical processes
- Review management structure
- Encourage effective employee participation.

**Quality**
- Satisfy customer needs
- Get close to customers
- Plan to do all jobs right first time
- Agree expected performance standards
- Implement company-wide improvement programme
- Measure performance
- Measure quality mismanagement and firefighting
- Demand continuous improvement
- Recognise achievements.

A second type of group training consists of creating a task force or ad hoc groups charged with solving specific but important problems. It is the responsibility of this team to manage any opposition to the introduction of new technologies, the accompanying training or the speeding up of decision-making processes.

The innovation contract is a third form of arrangement between the business and its employees, which enables the whole workforce to be involved in the most collective way possible. Many owners consider this kind of arrangement to be too restrictive, but it can certainly be harnessed to represent a real opportunity for all concerned.

In the case of the 460 companies in Lombardy, 30 per cent reported that they have permanent innovation contracts as well as consultation groups, with a further 15 per cent using this arrangement occasionally.

Naturally enough, the extent of participation varies depending on past experience, on the sector, on whether the culture is co-operative or conflictual, and on management behaviour. But it is the role of the initiator of innovation to assess these factors and to work towards a solution which takes them all into account.

Experience shows that there is little risk associated with these initiatives. What is highlighted again and again in these cases is that the workforce having prior knowledge of an innovation project, linked to an agreement covering its impact (for example, with regard to conditions of employment), never leads to a refusal of innovation, and frequently achieves substantially more than was initially hoped for from this type of co-operation. The figure below clearly demonstrates the diversity of possible solutions to be found within even a very small sample, involving the close examination of 21 companies undertaken in 1990 by the European Foundation for the Improvement of Living and Working Conditions in Dublin.

**Figure 6.6: Forms of Workforce Participation According to Innovation Phase in 21 Firms in Northern Ireland**

|  | Planning | Selection | Implementation |
|---|---|---|---|
| No participation | 5 | 3 | 0 |
| Information | 10 | 8 | 0 |
| Consultation | 4 | 8 | 11 |
| Contract | 2 | 2 | 4 |
| Joint decision-making | 0 | 0 | 6 |

## 6.4 Innovation and Training

What kind of training needs does a major innovation project generate? There are two key points at which significant training inputs are required: when the project is set up, and to ensure the successful management of innovation. Specific training needs are concentrated in three areas in particular: technology, organisation and human resources.

It is important to concentrate particularly on those forms of training which reduce the adoption and utilisation period to a minimum, and thus reduce both the resources used and the amount of work involved. Every technological project requires some form of training in which the employees involved will work with the producer of the technology to look at their particular function. Meetings between the suppliers of equipment and its eventual users should certainly be considered to be potentially extremely fruitful.

The smooth running of the new organisation depends on there being a clear fit between the new technological characteristics of the systems adopted. A flexible technology demands a modular or cellular organisation of work, whereas it will tend to be incompatible with a more rigid organisation. It is certainly true that the greater the decentralised and co-operative nature of the organisational model, the more likely the spontaneous involvement of the workforce.

It is equally clear that in those models which involve a highly-centralised organisation, and where everything is predetermined down to the last detail, the lack of free space and of relatively

independent business units will make the task of managing an
innovation project a relatively harder one.

### What Training Methods Are Available?

Although it is not the principal object of this book to do so (and
while accepting that the number of forms which training can take
is constantly changing and growing) this section covers three
widely used approaches.

**On-the-job training**. This is a permanent solution, although it is
often so unmethodical as to be almost invisible, even to the
trainee. The basis of this approach is that experienced employees
should take responsibility for new recruits. This approach to
training is often extended to include more systematic training activity, even in smaller organisations, and to the use of external
trainers who come into the business to deliver customised programmes.

The relationship between the cost and the rate of return on an
investment in training is obviously a key issue in this area for
any business which provides training. Yet there are a ways of
dealing with this:

- At the *business level*, the business can protect itself through
  an agreement with the employee to remain for a certain period
  after the training has been completed.

- At the *European level*, the European Union helps to check
  that the impact is lasting (European Social Fund and other
  projects in the area of training and retraining; exchange of
  employees between businesses and universities or technical
  schools).

**Training by local providers**. Even if, for the new recruit, the
on-the-job approach is a valuable introduction to the culture of
the business, it includes little which provides a preparation for
the process of change. It is very much in the interest of the business to use external training providers (chambers of commerce,
local Business Link, trade associations), all of which are capable
of providing useful but generally underused information on the
different opportunities which are available.

The available local and national training networks and re-sources remain underused by small and medium-sized enter-prises. The Formaper study in Lombardy, for instance, shows that the vast majority (81 per cent) use consultants, and that 14 per cent have internal counselling, but that only 4 per cent make consistent use of the range of external training organisations which nevertheless constitute the great majority of the profes-sional training pool in the region. This obviously raises questions about the quality and suitability of the training offered, but also about the way that it is promoted and the attitudes of business to training activity which may take place outside the business.

**Training from suppliers**. This type of training is indispensable. It is also probably the most easily accessible and the most rapidly effective. This training usually takes the form of an aftersales service linked to the acquisition and installation of equipment. To bring real benefits, this type of training requires a clear commit-ment on the part of the beneficiary as to the result to be achieved. This must be based upon the effective and lasting use of the new piece of machinery within the business as a whole; and not merely related to its use in technical terms alone. The failures to achieve these goals are frequent, particularly where office or workshop computer systems and equipment are concerned. It is too often the case that a machine which is sensationally efficient on the demonstration stand achieves little or nothing in a real working environment, either through inappropriate skills in the workforce, or through the equipment itself being poorly adapted to the tasks which it is required to undertake.

What usually occurs is that the supplier of the equipment trains the principal operator from the purchaser's organisation. It is then the latter's role to communicate this newly acquired knowledge throughout the organisation. But this is a much more difficult task than was the first. Although the distinction made above between first-hand and second-hand experience of training may give the impression of saving valuable time, it invariably turns out to be more expensive in the long run. This is particu-larly true when what is involved is the communication of an IT culture and mentality in a business with no previous experience of them.

The cost of this type of training is certainly high, but it is very much in the interest of the supplier to reduce the cost to the customer, since this is likely to lead to an increase in sales. And both parties have a vested interest in the equipment being used to its full potential, in particular because the sale of complex equipment usually implies a long-term relationship between the consumer and the provider of the technology. It is a relationship in which an imbalance can develop in favour of the provider, but it can equally be developed as a relationship able to provide long-term benefits for both parties.

### A Few General Issues Related to the Management of Training

To conclude this necessarily brief overview, it is worth recalling that it is essential to consider training as very much a strategic resource, which (just like R & D) constitutes a kind of investment in the future. Most businesses know the answer to the question "Which comes first, training or growth?" even if they do not always apply this to their own activities.

This is perhaps the right time to recall a few simple truths:

- The time which an employee spends in being trained is real working time, even if the return on this time (in accounting terms) is necessarily "deferred".

- There is a problem connected with rewarding training. Firstly, all successful training leads to an increase in the competitiveness of the business as a whole. It also increases the competence of employees, which itself needs to be rewarded. This recognition may take a monetary form, but can equally take other forms: recognition through promotion, through improved working hours, or inclusion in the Consultation Group, etc.

- Like the business as a whole, individual or group training has its own life cycle. Thus, at moments which are decisive in the life of the business (such as the acquisition of new machines or the introduction of new forms of organisation), the demand curve for intellectual or skills investment in the workforce will accelerate, and it is essential that the organisation's investment in appropriate training responds to this need. But equally, during the mature or even the decline phase of the life cycle of the machines, further investment in training in the technological and cultural areas or in that of general management can lead to a revitalisation of the technological life cycle.

- Training is inextricably linked to the development of employee involvement in the organisation as a whole. Thus, independently of its obligatory character (as is the case in France), it must be seen as a specific strategic tool, a key resource in the business as a whole.

- Training plan and business plan develop and change together. This idea finds its formal development in England and Scotland in the Investors in People national standard.

### Figure 6.7: An Investor in People

---

- Makes a public commitment from the top to develop all employees to achieve its business objectives

- Regularly reviews the training and development needs of all employees

- Takes action to train and develop individuals on recruitment and throughout their employment

- Evaluates the investment in training and development to assess achievement and improve future effectiveness.

---

The available evidence increasingly suggests that this form of investment provides real benefits in terms of improved customer service, quality, staff retention and profitability.

# Chapter 7

# SETTING UP THE INVESTMENT

*The next stage can really be described as "making it happen". It is the beginning of implementing the innovation process that has been selected. Whereas the decision-making phase drew a mass of thought and actions towards a single point, this stage involves the reallocation of tasks. The first step will be to create a work programme, to be followed by putting in place the materials, carrying out trials, and finally beginning production.*

*Starting up and increasing production is certain to raise a number of problems, whose resolution will not always have been anticipated and which can be expensive in a variety of ways: in overtime, in time spent, in delays for the customer and in reduced production. These unexpected delays can be contained, however, through a realistic and flexible preparation of this stage of the process. This involves anticipating the ways in which the change will be carried out and involving from the outset the future users of the system in planning its installation. There are two objectives to this approach: to stick to the planned time-scale and to reduce incidents as much as is possible. If these objectives are to be achieved, then four conditions must be met:*

- *A rigorous preparation of the technical elements, containing detail both of the materials and their environment (position, safety, maintenance, arrangement of work stations, ergonomics).*

- *A genuine involvement of everybody who will have to make the new system work; and a broad consultation having taken place with regard to the most appropriate organisation of work (qualifications, recruitment, training).*

- *Detailed development of logistics (flows of materials and flows of information).*

- *Most important of all, effective human and technical co-ordination, either by an individual or by the team which is specifically responsible for carrying out this phase.*

## 7.1 Planning the Setting-up

The "making it happen" phase can be broken down into two parts: that in which the material or the new computer system is installed and that in which it is launched. The key moment in the implementation of the project is when tests take place both for the effectiveness of the methods chosen and for the accuracy of earlier studies and decisions. The most frequently encountered difficulties tend to occur for the following reasons:

- The technical capability of the machines or of the system has been overestimated, or insufficient allowance has been made for the constraints which everyday use brings.

- The importance of human factors and of organisational problems have, on the other hand, been underestimated: the impact of change on the organisation and on its working practices have not accurately been measured.

The sheer number of potential projects makes exhaustive description impossible. This chapter will focus therefore on an analysis of a number of methods and the underlying principles. But even if it is not possible to go into much practical detail here, reference will continue to be made both to industrial projects, to innovations in management practice and to market innovation.

A few general remarks can however be made based upon experience of the kind of problems that often occur:

- The way in which this delicate period is managed has a major impact on the eventual success or otherwise of the change process. Uncertainty has now moved to the other side of the equation. A decision has been made as to the overall direction to be followed. The objectives are clear and substantial resources have been allocated. But these decisions have a large and direct impact on many aspects of the daily life of each employee. Major strategies often ignore the daily concerns of others, and major ambitions are often brought down by minor details. The difficulties which can be anticipated create a sense of tension which can cut both ways: if it is kept within limits, it can encourage individuals to exceed expectations; but, beyond a certain point, it can degenerate into conflict, aggression or even into giving up. Management by stress can cause guilt, or by fear of error and punishment can paralyse, at the very point in the process where every individual needs to work differently and to take initiatives to overcome problems. It is thus the role of management to identify an acceptable level of stress and to control potential sources of losing control.

- It is during testing that the first incompatibilities between the different pieces of equipment, between the equipment and the material, and between the equipment and the people, are discovered. These setbacks will cause everybody problems, in the same way that somebody using a newly-acquired computer programme only discovers through using it that it is inappropriate. In many cases, subsequent adjustments or changes carried out by specialists can lead to the initial cost of purchase being doubled. There are no shortage of subcontractors, encouraged by their principal customer to take steps to ensure the exchange of computerised data, who have run into the incompatibilities between the two networks. In most cases, this sort of incident can be eliminated through a more thorough initial study of the choice of equipment and in particular through using specialist advisors.

- Loss of machine time due to repetitive breakdowns is a major factor in time-scales not being respected. They cause major problems to the marketing function and often lead to subcontractors having to be used.

- Quality problems in equipment, in new inputs or in the linking elements may only become apparent during testing and often reveal inadequate flows upstream of this operation. This is frequently due to discussion with suppliers not having taken place early enough in the preparatory phase to be able to obtain a reliable product which is appropriate for the new process. The development of partnership with suppliers often happens too late.

- Finally, in strictly technical terms, implementation will often bring new difficulties to the surface: difficulty getting access to key parts of the machine to carry out maintenance, overload at one work station compared to the others, absent or poorly-adapted control procedures.

These kinds of disruptive factors do not of course always crop up, and certainly never occur in the order one would expect. But they have been described so as to underline the potential benefits of a rigorously-conducted installation phase.

At this stage, the tasks to be carried out can be broken down as follows: installing the equipment; testing and checking; training the users; developing the upstream and downstream software. Both the installation and testing, as well as the training and development, need to be carried out in parallel so as to be able to draw together, when they are completed, all the factors which contribute to launching. Thus it is necessary to make simultaneous progress in the areas of involving the workforce, planning operations and assembling the technical files.

### Integrating the Project and the Workforce
This issue is considered here only in the context of the installation phase. The following need to be known:

- Who should participate in the process?

- When?

- In what role?

Integration needs to involve all those who are going to go into the business, even if only for a short while, with a view to carrying out the installation and the initial usage of the innovation. These people possess problem-solving skills and action skills rather than theoretical knowledge. The object is thus to organise the transfer of their knowledge to the employees of the business. This integration needs to be both logical and gradual, so as to ensure that future users have an organised appreciation both of the project and of its development. Any involvement of the right person at the right moment needs to be planned, inasmuch as this is possible.

This will certainly lead to the theoretical vision of its creators having to confront the day-to-day experience of the users. The latter, in an industrial project, will be able to give feedback on the reaction of different materials, on the nature of the most frequent breakdowns, on the best way of solving problems, on the ways in which the manual should be used, and on the practical problems associated with handling in general. All this information makes it easier to predict (and therefore to plan) how the start-up should happen, how the new equipment will function, and how maintenance should be organised.

Discussion on technical questions with those responsible for the project helps prepare the users for operating the new equipment. Little by little, they create new ways of understanding and visualising what they are doing, of shaping their knowledge and of making the new techniques their own.

However, this two-way exchange of information can only bear fruit if two conditions are fulfilled:

- The powers and freedom of action of the implementation team must not be restricted for any reasons of internal hierarchy.

- Individual roles must not be too rigidly defined.

**During the initial studies,** discussion needs to have taken place, even before the investment decision is made and the technical choices become irreversible, between the creators of the project and the operators, a few managers and the supervisors. In an industrial project, this discussion can include, for instance, a manufacturing engineer assisted by one or two experienced supervisors. Operators can similarly be consulted on certain questions. This is one of the aspects of the consultation process described in Chapter 4.

*In the course of more detailed studies,* a number of decisions are therefore already irreversible. The object of the consultation which takes place is to prepare the actions which need to be carried out prior to the installation of the equipment, such as searching for suppliers, issuing tenders, recruiting staff. This group is thus a more diverse one. The operators, who are the future users, are the designers, the workshop managers, the departmental mangers and the general mangers. Anybody who will have any involvement on the ground must be involved, in some form or other, at this stage. Their participation draws not only on their experience; it also mentally prepares them for the launch. As the frequency of the periodic co-ordination and progress meetings increases, the impact of this on the workload of the participants means that their replacement in their current job and other reductions in their current activity all need to be anticipated.

Using this participative approach involves immediate indirect costs (see Chapter 8). It also changes behaviour, since it has a disruptive effect in the way that it temporarily removes the hierarchical nature of working relationships and the barriers that exist between different areas. In the longer term, however, it can bring real benefits through enabling the participants to experience a more flexible and open means of operating and communicating, much of which they will naturally retain when they return to their former role.

*At the point at which the equipment is finally installed,* the team will have grown due to arrivals from outside: suppliers of equipment, electrical and plumbing contractors, etc. Two additional types of person are also likely to be highly visible on site: health and safety officers, who are checking that all the necessary steps involving and around the installation of the equipment are being taken; and the future operators, who are present to assist with setting up the core equipment and the various related elements. Those who will be responsible for its maintenance have a particular interest in checking it out when it is stripped down and in developing a real grasp of the technical drawings through a process of observation.

### Planning the Operations

The planning process is intended to organise the way in which the installation is carried out, so as to co-ordinate the different operations and leave nothing out. It consists of the following:

- Making a list of the different stages which have to take place

- Deciding, within each stage, on the order in which the different tasks need to be carried out

- Anticipating all the activities which are related to the installation and have to be carried out in parallel

- Drawing up a plan which ensures that everything can be carried out in time.

The planning process leads to two other kinds of consultation. The first takes place inside the group of those involved in the project, as described above. The second takes place between employees from the business and the representatives of external organisations who are involved in the project: the suppliers of the equipment and those who will install all the related equipment and carry out work such as making changes to the workplace, electrical work, etc. Those businesses which work with a large number of external specialists always run the risk of seeing the work broken up into tiny pieces, and a lack of co-ordination and technical adaptation leading to a number of delays with all the knock-on disruption that this can cause. To guard against these difficulties, any work plan needs to take place in consultation with both the installer and the customer.

This planning involves establishing a technical organigram for ordering tasks, starting with a breakdown of the elements required to several levels, depending on the amount of detail that is appropriate for a particular project. Planning the operations and the time involved can take different forms, the most frequently used being the Gantt diagram, which is simple and easily read, or a chart drawn up using the PERT method. Both of these are described below.

**The Gantt Diagram**. Henry Gantt, an American engineer who was a specialist in the scientific organisation of work, devised a diagram designed to organise work or tasks within the following three constraints:

- Lead time

- The resources required

- Logical links with other operations in the plan.

The approach consists of defining the sequence and overlap of the work and of positioning it on different work stations taking technical restraints and throughput into account. Where several different combinations are possible, the decision is made based on the experience of the person who is doing the planning and on a few simple production scheduling rules:

- Priority is given to the earliest delivery date

- Priority is given to the shortest operation

- Priority is given to the smallest gap between the time left before completion is required and the amount of work still to be done.

**The PERT (Programme Evaluation and Review Technique) method**. This method achieves the same ends as the Gantt diagram and was first used by the US Navy during World War Two, when it was developing its nuclear strike force and needed to co-ordinate the activity of 250 major suppliers and of more than 9,000 subcontractors.

In the PERT method, each operation is represented by an arrow on the diagram and takes place between two staging posts, each of which is represented by a circle. This assembly of operations and stages creates a network on which the duration of different tasks is recorded as well as the order in which they need to take place.

The calculation of the duration of each operation and of the total time is made up of an average taking into account an optimistic time, a pessimistic time and a realistic time. (The final plan consists of two timescales: that of the earliest possible completion date and that of the latest possible completion date.) The difference between these dates gives a safety margin which enables delays in certain tasks to be tolerated without the completion date of the project needing to be altered. The chain of tasks for which there is no margin, on the other hand, represents the critical path upon which management attention should urgently be focused, and action taken to ensure that the critical tasks receive all the resources they require.

This diagram is put together as a result of preparation work consisting of:

- A list of the different stages

- The definition of the different tasks

- Designing the network

- Estimating the time required

- Calculating the critical path

- Making adjustments to the project as a result of this

- Circulating these results.

Carrying out the project needs to involve a constant monitoring of progress against plan in all areas, by a comparison of real time to estimated time and in an adjustment of the planning schedule in the light of experience.

The real merit of this method is that it defines responsibilities in advance, communicates information to everybody involved and provides a permanent stimulus to the users.

**Managing the site as a whole**. Installing new equipment or computerising a process can only take place if everything into which they link is ready to be used in normal working conditions. So it is important that any foreseeable changes which need to be made in the organisation of material flows and the transmission of information should be carried out at the same time.

The overall management of the installation comprises:

1. The environment for the equipment
   - Space
   - Health and Safety
   - Maintenance.

2. The organisation of work
   - Employment and management of staff
   - Work stations
   - Qualifications and training.

3. Flows
   - Information flows between workshops, stores and administration
   - Material flows (equipment, products).

Each of these roles is allocated to a different person and requires a detailed study and the assembly of technical files. All this needs to be co-ordinated within the project engineering team.

**Managing the unexpected.** This too needs to be planned for. This statement may appear to be a paradox, because the unexpected is, by its very nature, beyond definition. Nevertheless, the very fact that it is considered is a way of preparing to deal with it. It is the sign of a certain openness of mind, and of tolerance of the kind of mistakes everybody makes at some time or other. It also shows an awareness that one can never have all the possible information at one's disposal. In short, a modest attitude when faced with complexity places the participants in a better position to resolve the problems.

Several methods are possible. The first is to anticipate, in the installation plan and for each stage of the process, the need for a *contingency plan* which describes a course of action.

The *scenario method* can also be used to reduce the impact of the unexpected. This consists of imagining all the situations which could arise and of inventing a reaction to them. The question is always the same: "What if . . . ?" From this point the following stages need to be carried out:

- A risk analysis
- A detailed study of technical information and trouble-shooting instructions
- Drawing a tree of causes.

A list of the principal breakdowns needs to be drawn up, with, for each of them, the appropriate response. These fallback solutions, listed and then distributed to all concerned, form an emergency plan which can also be used in training exercises involving an imagined failure. This plan will be even more effective where the analysis and the report are carried out by the operators themselves, under the guidance of the engineers. Experience tends to show that this has been anything but wasted time.

Those individuals who have most demonstrated their ability to cope with the new situation should be placed at key points in the process, since the role and the quality of middle management is crucial at this stage. Since it is never possible to anticipate every potential area of breakdown, the aim should be to prepare the operators to take the right steps. They need to be able to think through the methods (and not the solutions) which have been used in similar cases and to devise appropriate responses from these. It is thus through using an approach based on reason, and not merely a troubleshooting checklist, that these key people need to be trained.

**Defining the evaluation criteria**. This should also be carried out at the point at which detailed studies are undertaken. In this case, and for new technologies, the criteria are not only the manufacturing costs (see Chapter 8). The pursuit of productivity and quality means that a choice has to be made as to the indicators of operational performance, rather than purely of return in quantitative terms. Only two or three are needed, as long as they are well chosen.

For example:

- Delays against the plan
- Unbudgeted overspend, which indicates a planning error and whose origin it is necessary to identify (waiting for the sub-

contractor to come and deal with a problem, lost production, etc.).


## Putting Together the Technical Files

It is worth taking a closer look at the technical files. Each project requires two technical documents, the **master plan** and the **terms of reference**. As was described earlier, they consist (as well as the guide for installing key equipment) of appropriate files on the subjects of preparing the site, security, maintenance, the organisation of work, and the handling of material and immaterial flows. Each of these files must be the responsibility of a specialist in the particular area. The management of any kind of flow demands a particular attention because of its impact on achieving the strategic objective of quality.

The management of components in large quantities (parts, sub-components, supplies) creates a need for contracts with suppliers or subcontractors. Their production is the opportunity to undertake preliminary studies with a view to ensuring quality and to establishing regular checks. Those which are produced within the organisation need to be subjected to time planning, linked to supply requirements. During the test phase, it is wise to anticipate over-supply.

The terms of reference which go with this consists of a parts list and of plans which must, in all cases, be identified by a double code: a technical code to indicate where the components come from, and a functional code which indicates their role in the installation as a whole. Their handling can thus be computerised, allowing a selection according to type of material, function, system, etc.

The management of the technical documents is, given their quantity, extremely important. Lessors of licences, suppliers of equipment, R&D workshops, external specialists, for instance, will all produce contributions which will need to be used by people who have not taken part in the elaboration of the project. For this reason, the creators of the project also need clearly to classify these.

Certain documents are put together specifically for the putting into operation of equipment, and to be used at later stages in its life cycle: a record of receipt, an activity sheet, etc. These make up the technical history to which it is essential to refer should any problem occur. They can also provide a useful source of reference for subsequent projects, which is why it is important to record details of incidents which occur during the setting-up period, as well as the way in which they have been resolved.

Obtaining optimal use from new equipment thus depends on the linkage and coherence of all the internal departments and external networks involved. It is clear therefore that only an integrated logistical approach can limit the inefficiency and wastage which an ill-adapted, fragmented approach would be certain to bring.

## 7.2 The Launch

Two particular stages demand brief discussion: the trial period and start-up.

### *Trials*

It is usual to distinguish between the trials which are carried out by the installer and those which are carried out by the representatives of the customer, who will be the operators of the equipment.

It is normally only after the trial period that the transfer of responsibility from vendor to customer takes place. The receipt period thus often reveals the conflicting interests of the installer and the customer. The former wishes to accelerate the process so as to be paid as soon as possible, and in cases where there is a problem to invoice subsequent involvement on top of the cost of purchase. The customer's interest, on the other hand, is to attempt to drag out the process for as long as possible so as to ensure that the installer bears the cost of initial teething troubles.

So it is prudent to define precisely, in advance and in a written contract, the responsibilities of the provider, and to demand certain guaranteed results in the pre-launch tests. In most cases, a little extra expenditure to pay for the supervision of the operators within the business during the first tests often leads to savings in time and money over the operation as a whole. Certainly the potential for conflict will be reduced if the business has taken steps to make the handover of responsibility as smooth as possible, through taking care to ensure that its employees have been fully trained in all areas.

### Starting up

New problems often appear at this point. Delays at launch are very common and are often of several months, or even more. The most frequent cause is generally problems which have not been solved at an earlier stage. The length of the delays is thus a key indicator of non-performance.

Once the installer has left, the business discovers the problems associated with everyday usage: throughput reaching saturation point, unsuitable operation and storage of data on disk, or errors in the system when orders occur too close to each other in time.

Nobody can eliminate this type of problem altogether. Nevertheless, limiting them to an acceptable level depends once again, at least in part, on the use of a logical and systematic procedure as well as the involvement of technical experts, regular and thorough checks, and, where the workforce is concerned, increasing (in the short term) the size of the operating teams.

To sum up, it is clear that time intelligently wasted during the preparation phase is always time gained, at the very least, at the point of launch. Successful transitions from old to new technologies are made by those organisations which closely relate the pursuit of technical improvement to the involvement of all their employees.

# Chapter 8

# EVALUATING THE COSTS OF THE PROJECT

*Before choosing a project, it is important to have come up with a clear estimation of the costs, divided into those which are certain, those which are probable and those which are possible, together with the periods in which they will be incurred. This makes it possible to calculate the cost effectiveness of each proposed project as well as its capacity to meet the industrial and commercial objectives which have already been defined. The standard costs of production, which are established from the provisional budgets, are a key element of this.*

*This initial estimation is also the best way of controlling, throughout the period of setting up the project and also during its operation, that the results are in line with the objectives. Any variations which appear between the standard costs and the real costs need to be looked at very closely, so that remedial steps can be taken. Chapter 9 will look in greater detail at the impact study, that is to say the final results.*

*But control of the costs, however well adapted it may be to a standardised system of mass production, is not normally enough where new technologies and the innovation process are concerned. Strategic control of the activity is not merely aimed at achieving economies of scale, but also at measuring and improving levels of quality, speed of reaction to demand, and efficiency in production methods. These costs cannot be measured in quantitative terms alone.*

*If the reduction of costs is to be viewed as part of performance, any reduction in performance also has a cost.*

*This is why project managers need to establish, from the moment at which new equipment or new material is introduced (and thus before it starts operating), what the indicators will be against which its productivity is to be measured. These indicators need to be outside the normal accounting processes, such as, for example, the time which it takes a team to produce an order, the length of a production cycle, the time it takes to adjust a machine, the rate of return by product type, etc. These indicators also have a significant motivating role in the way in which they focus the operators on the key objectives.*

*The process of choosing these indicators is an important act of strategic control which concerns everyone involved in the project.*

*So, alongside the quantitative indicators (time spent, or resources used), qualitative indicators have an equal impact on any assessment of a project's success. This is true in the case of the acquisition of specific know-how along the learning curve, of employee resistance to the change process (or, on the other hand, of employee collaboration). It applies equally to the ease of reaction to problems and unexpected mishaps, to the way in which new procedures are embedded, to customer loyalty, and to the reputation of the business.*

## 8.1  The Aims of Evaluation

The importance of evaluation has now been raised at a number of points: it is important during the Diagnosis, in order to measure the internal resources of the organisation; and it has also been used to clarify the key factors in a choice of strategy. The process of evaluation starts therefore as soon as one begins to ask questions about the development of the business. It continues throughout and beyond the implementation of the project.

Nothing prevents a business from investing before it has assessed the risks involved. This may appear an unwise approach, but it is no more unwise than failing to monitor performance either during the period in which an innovation is put in place, or during the period in which the anticipated output is due. The same is equally true of the final analysis of a project. Every single element of feedback and discussion informs the learning curve of the creators and operators of a project, and adds to their competence to make the right decisions in the future. Evaluation is thus, at the same time, a tool for analysis and diagnosis, and a means both of control and of strategic guidance.

**Figure 8.1: The Cost of Change (the Later the Change, the Higher the Cost)**

| Stage | Relative Cost |
|---|---|
| Concept | 1 |
| Detail design | 10 |
| Tooling | 100 |
| Testing | 1,000 |
| Post-release | 10,000 |

## A Method for Analysis and Diagnosis

Evaluation has at least two vital functions. The most immediately felt is in the area of *finance*. Faced with several potential projects, which is likely to the most profitable? Is it able to be self-financing, and how long will this take? What capacity does the business have to finance the project itself, and to what extent is it able to borrow?

Any investment needs not only to strengthen the market position of the business, but also to secure its future. Before making a decision, therefore, it is necessary to measure, using all the data that is available at that point, what its impact on overall activity and productivity is likely to be, so as to be certain that it has every chance of achieving its objectives. This is *evaluation* in the sense of a financial calculation, which focuses specifically on the costs of production: it enables the measurement of any economies which are achieved due to the use of the new equipment, and informs a study of the impact of this on costs. Provisional budgets (both for investment and for production) are drawn up. The data is provided by financial accounting and cost accounting.

*An Instrument for Measuring Impact*

The data used in forward planning (price of materials, depreciation of the equipment, inputs, employee costs) enable the calculation of the standard production costs which serve as the reference.

These are the average or provisional costs, defined in terms of the unit of production which the business has chosen. They include a range of different kinds of cost: fixed overheads and variable overheads, charges which are directly related to the product, as well as indirect costs allocated by means of common units of measure. This last type of cost needs to be spread between the different cost centres, in the usual way that the business makes this kind of allocation. The standard costs, once they have been calculated using this method, are thus also related to actual production and outputs.

When the results are analysed after the event, the variations with the provisional calculations will emerge. These need to be closely examined, and any necessary adjustments made.

The task of collecting and processing the quantitative data relating to the production process belongs to the accounts department and to the specialists in budgetary control, who are thus able to provide, through their chart of the key business indicators, the variances under the following headings: materials, labour, overheads, output (or production), return. The interpretation of these results, on the other hand, needs to be carried out by an analysis session that involves the key people in each of the different areas. The factor which is the source of the variation (for instance, cost of materials, or slow stock turn), or is identified to be the cause of a fall in output, can only be highlighted and explored by those who are involved in the project on the ground (that is to say, the technicians, marketers and administrators) in collaboration with the planning team and perhaps the management consultants.

Any in-depth evaluation of a project thus involves everyone in the organisation who has had any kind of role in putting it into place. It should not focus exclusively on financial data and indicators, but also consider technical issues and their impact in the market. Evaluation needs to take place, therefore, using the reconstituted original Consultation Group.

## A *Tool for Steering Strategy*

Up to this point, evaluation has relied principally on the traditional forms of budgetary monitoring and control. It is important to know for certain whether these indicators really are achieving all their expected objectives. Management by production costs alone is never entirely appropriate where the strategic management of new technologies is concerned.

**The limitations and perverse effects of the traditional ways of calculating costs**. Experience proves that, however indispensable the calculation and measurement of production costs may be, it nevertheless has significant limitations:

- Standard costs that are calculated in relation to cost centres or other traditional methods are based upon a particular view of the structure of the business and of the rules for allocation that senior managers have defined. They are thus based upon convention, rather than having any inherent value. Any major innovation has the effect of substantially altering the general shape of the business, and so should cause these conventional assumptions to be revised.

- The nature of the costs changes as a result of the impact of change on the organisation, and in particular where the qualifications and job descriptions of the operators are concerned. Direct costs are the decisive factors in most forms of mass production, but in this new situation they are reduced, on the whole, to materials and to specific investments. The labour cost, on the other hand, becomes an increasingly important element of the indirect costs, and so must be added to the costs of depreciation of other investments, and to the costs of non-quality, general administration, storage and maintenance. So the principle of direct cost has less and less relevance in a business which has adopted an innovative strategy that includes factors such as customer service and customer loyalty.

- The same holds true for the relationship between fixed and variable costs. The automatisation of production reduces the share of short-term variable costs such as materials, energy and semi-processed goods. As for the so-called fixed costs, they do not increase with production volumes, but vary solely due to the complexity of the activity and to process enrichment.

*   The rigid use of costs as a measure of efficiency can have a perverse impact when it comes to managing an objective based upon quality or flexibility. Through asking a workshop to respect a standard cost based on a certain level of activity, one effectively encourages an increase in production. This results in a flood of work in progress at the very moment when a key objective is the reduction of stock levels. In this case, therefore, there is a clear contradiction between the standards imposed on the operators and the overall objectives which the business has established for its own growth.

**New costs need to be identified by the management information system**. The traditional way of calculating costs fails to include certain elements which nevertheless have a real impact on the final outcome of any operation. The requirements of accounting legislation involve the inclusion of visible, named, measurable, monitored costs. But they neglect all those costs which are the result of the unmeasured cost of poor integration, as well as all those which are deliberately accepted with a view to achieving improved performance, such as hidden costs and opportunity costs.

*Hidden costs* also manage to escape the normal information-gathering process. Some are mixed in with the visible costs without there being a clear link with the desired objective (for instance, paying for the temporary staff who take the place of absent full time staff), while others represent a loss of value, as in the case of work time which is paid for without any resulting production (the salary of an absent member of staff, for example).

*Opportunity costs* are due, for instance, to keeping a machine idle so as not to have to replenish stocks, to stopping a machine to carry out preventative maintenance, or through slowing down production to ensure quality. The pursuit of total quality, of total productive maintenance, or of just-in-time, requires a different approach to calculating these costs which all contribute to effectiveness.

The **costs of not having quality** (sometimes known as the cost of non-conformance) are invariably far higher than is thought. These costs can be categorised under four headings:

**Cost of prevention** (stopping any deviations from specification):

- Engineering the product so that it can be put together incorrectly
- Checking product specifications and drawings
- Preventative maintenance of process equipment
- Developing and operating quality-measuring equipment
- Administering quality procedures
- Surveying quality levels, problem-solving and implementing quality improvement
- Supplier appraisals and training
- Training and development of staff.

**Costs of appraisal** (checking to see whether any deviations from specification have occurred):

- Product prototype testing
- Inspection and test of incoming goods
- Inspection and test of internal processes
- Field checks on product performance
- Processing inspection and test data.

**Costs of internal failure** (coping with non-conformance inside the organisation):

- Scrapped parts and materials
- Reworked parts and materials
- Diagnostics and quality defects and failures
- Lost production while process is stopped
- Reorganising processes and procedures after failure

- Lack of managerial concentration and focus caused by trouble-shooting rather than searching for improvement.

*Cost of external failure* (the cost of the product failing after being handed over to the customer):

- Warranty costs

- Servicing costs

- Product liability

- Complaints administration

- Loss of customer goodwill affecting future business.

Up to this point this chapter has concentrated on those factors which best lend themselves, more or less directly, to some form of quantitative evaluation. There are also a number of other factors, however, which can only be detected through a qualitative approach. It is only through listing them, and ranking them in order of importance, that one can begin to appreciate their number and effect. A number of these are to be found in the box below, but this is in no way an exhaustive list. As will be seen, these factors are anything but negligible. They may start off as the narrow winner in a decision between two solutions whose calculable costs are very close to each other, but they can nevertheless become factors which prove to be decisive in the attempt to enter and establish a presence in a particular market.

**Figure 8.2: The Three Levels of Evaluation**

Strategic management by activity places considerable importance on the way in which a product is produced. The production cycle is broken down into processes and activities. The next step is to evaluate the three levels at which overspends which need to be monitored occur, and at which potential improvements need therefore to be identified and carried out.

## 1. Quantitative factors which are directly calculable

What is required here is a medium-term evaluation (from 18 months to 2 years) which highlights the requirements without which a project does not get off the ground.

*Additional costs* are due to:

- An overestimation of the total capacity required in order to satisfy maximum demand; to the decision to acquire multi-purpose equipment which is more expensive, more fragile and therefore less productive; to achieving a higher skill level among the operators.

- A loss of capacity due to the adjustments involved in the frequent changes of production run; to higher reject rates; to the workforce being less productive; to large amounts of handling.

- An escalation in the number of small-scale deliveries, resulting in an increase in packaging and transport costs.

*Potential savings* resulting in:

- An increase in the rate of return

- Improvements in stock levels and in factor utilisation, lower use of energy, of materials and of semi-processed goods

- The rapid detection of defects and the reduction in the number of faulty items, and adjustments to machinery carried out rapidly due to experience

- A reduction in direct labour.

## 2. Quantitative factors which cannot immediately be calculated

The area under examination is extended beyond the workshops to the administration and management functions. The time-frame under analysis is also extended, to cover the working life of the equipment.

The ***additional costs*** are due to:

- The fragility of the new system (machine stoppage time) and to the reduction in overall productivity due to the impact of this on all other service areas

- Having to call in temporary indirect labour, which is less productive, unlikely to stay for very long and insufficiently qualified

- An overspend on direct labour due to flexible working arrangements (teams on 3 x 8-hour shifts, as well as the weekends)

- The workforce being either over-qualified or under-qualified.

***Potential savings*** due to:

- Being increasingly competitive in the marketplace (broadening and diversification of the customer base; increased customer loyalty)

- Being able to get hold of a higher quality of stock from suppliers

- Lower costs for stock control and accounting

- Improved working between different areas of the business due to their increased integration.

### 3. The qualitative elements

The longer term impact is to be seen in the overall performance of the business and in its ability to achieve excellence.

The ***additional costs*** are due to:

- The loss of industrial know-how which may be outdated but is nevertheless rich in possibilities

- Employee resistance to change

- Difficulties adapting which have either not been identified or not been ironed out through training.

---

*Potential savings* resulting:

- From changes in the overall objectives of the business, and improvements both to motivation and to internal relations

- From going up the learning curve, and from the development both of a culture which is adept at change, and of the know-how to solve problems

- From more integrated management

- From speeding up the decision-making processes

- From the embedding of procedures.

---

**Impact of costs on the rest of the business or on indirect costs**. Any major innovation always has a knock-on effect on the technical conditions of production and on the organisation of the marketing and general administrative functions. This effect consists of the following:

- Those flows linked to the new equipment: information flows, upstream and downstream material flows

- The way in which work is organised: job descriptions, organisation of work schedules, new qualifications and multi-skilling, the composition of different teams

- Monitoring the product life cycle and its renewal, which has repercussions in the areas of marketing and administration

- The indirect costs of maintenance, security, and subsequent restructuring.

In the majority of cases, these indirect costs cannot be allocated to the final cost of the investment, nor included in the financial plan, outside the standard calculation of a margin of security. But they are part of the cost of the project and will certainly have an influence on its eventual profitability.

This section has covered most of the elements which make up the true price that is paid to achieve new goals such as quality, productivity and responsiveness to demand. When they are included in the overall picture, they reveal **not the cost of production, but the true cost of a decision**.

Nor should one stop there. For, once they have been identified, these costs need to be examined and reduced. It thus becomes necessary, in a second phase of activity, to find new ways of controlling hidden costs. This is the role for new performance indicators.

## Figure 8.3: The Meaning of Words

---

### What is activity?

Activity is everything in the life of a business that can be described using verbs: to design, to mix, to assemble, to prepare a budget, to send out invoices, to visit a customer, to give preferred status to a supplier.

It is a combination of simple tasks:

- Carried out by an individual or a group

- Making use of specific know-how

- Linked in terms of their performance as regards cost and result

- Enabling an output of some kind (a component, or a budget) to be produced for an internal or external customer, using a combination of inputs (labour, equipment, information).

### Identifying the significant processes

A process is a sequence of activities which are spread over time and which result in a common objective. For example, the preparation of a plan, the sanction of a supplier or the calculation of a production cost.

The underlying idea is that performance is built out of what is done, and by the way in which it is carried out. It is not simply a question of drawing up an exhaustive list of business activities, but of selecting those which are most significant for the chosen objective.

---

**Operational indicators**

These are the indicators of productivity, of capacity for reaction or of quality which make it possible to be able to work back to the root causes of poor integration and of overspends. On this subject Ono, the founder of the Toyota management system, recommends that the question "Why?" should be asked five times, so as to avoid being satisfied by the first visible cause when the true cause lies somewhat deeper.

Among the operational indicators are to be found the *key performance indicators,* which are designed to guide an action currently being undertaken (the number of approved suppliers, the length of the production cycle, the reject rate or the time it takes to process an order) and the *strategic indicators,* which are uses to monitor the coherence of activities rather than to ensure hierarchical control (the time it takes to change tools).

## 8.2 Strategic Management

The aim of this approach is to focus everyone in the business on the new factors which have an influence over performance; that is to say, to produce with the maximum possible operational efficiency.

In the same way that a very slight correction in its course is sufficient to return a satellite to its orbit, the introduction into day-to-day management activity of simple performance indicators linked to the most frequently-performed activities can have a decisive impact on the way in which individuals adjust their behaviour to achieve strategic objectives.

These indicators are essentially data from outside the usual accounting circuit. They need to be defined prior to new equipment being used, and selected as a result of consultation between the initiators of the project and those who will be using the equipment. The indicators need to cover all the different areas of the business, including technicians, marketers, administrators and managers. Additionally, they need to be transparent and accessible to all future users, with regard to coherence and scope.

An example of this could be the cost of an electric motor in a goods lift. This is a semi-processed element, a sub-assembly which is integrated into a final product. The cost of this motor includes the price paid to its supplier and the cost of storing it. In management by activity, the aim is to calculate the exact cost of each operation in the motor handling process. A list is made of the different activities which are carried out in the warehouse: receipt of the delivery, checking each element (whether the motor works, but also the state of the packaging, of the paint work, etc.) and eventually even returning it to the supplier, following up the return, and supplying the workshops. The cost of each activity is calculated by allocating a number of units of time, to which is added the depreciation of the equipment used, and any other costs.

Any analysis of these costs concentrates on looking to see what savings can be made on the activity itself, through eliminating unnecessary operations, or introducing new forms of quality control, etc. The object is to work back to discover the causes. In order to reduce the quantities which are returned to the supplier, it is not just a question of selecting those suppliers which are the most reliable but also of working to develop partnership with them to ensure that this high quality does not deviate in the future (this is the process of approved-supplier status).

It is in this way that the performance indicators are able to have an impact which results in higher quality and optimal speed of reaction and flexibility. They make it necessary to track down useless operations, which develop rapidly where there is either routine or carelessness, and also to identify those operations which could more effectively be automated or carried out by computer.

To select the indicators, it is necessary to begin with an *analysis of performance* around the following questions:

1. What improvement is required?

2. How can one obtain it?

3. What is the criterion for achieving it?

4. What information can measure it?

Let us take the example of a flexible bottling machine which allows different combinations of contents (yoghurt, fizzy drinks, etc.) and containers (pots, bottles, etc.). The performance of this piece of machinery can be evaluated using the four questions.

The first question defines the *objective*: to increase the machine's capacity to follow changes in market demand.

The second question identifies the *process* which is the lever of the activity: shortening the time taken in the manufacturing cycle, of which the current inertia slows down the whole production system.

The third question focuses on the *critical activity*: the more rapid management of stocks of finished product, of work-in-progress and of raw materials, as well as the rate of delivery by the suppliers. Because all the reasons for maintaining high stock levels have to be eliminated, this activity needs also to look at failures to maintain quality, at machine breakdowns and at the time it takes to set up a machine.

The fourth question defines the *performance indicators*: volumes of stock, the rate of stock turn, the time it takes to change a tool and to set up the machines, the machine down-time for maintenance and cleaning, the number of breakdowns, the number of supplier returns.

Each of these indicators implies a change in behaviour by the users who are touched by it, and a need to be attentive with regard to quality rather than simply to output; in short, they need to think about the whole rather than merely their part of the process. These will lead, over time, to other changes, such as a rearrangement of storage space, the setting up of procedures to establish preferred-supplier status, new training programmes for the warehouse staff and other team members who are implicated in the operational effectiveness of the whole.

The introduction of this form of evaluation should only be gradual, needs to be carefully managed, and is best limited to the specific project. Initially, a small number of indicators will certainly be sufficient: not more than five or six per project. Where possible, they should be able to be established using data which is already collected and available, so as not to cause resistance by increasing the workload. These indicators do not automatically replace the units of measurement which are already in use, but they do need to link logically into them. For example, it is important to eliminate any contradiction between an objective of improved quality and a rewards structure based solely on output. There only needs to be one objective indicator: in the above example, it could be the length of the manufacturing cycle.

To sum up, therefore, it needs to be recalled that the strategic steering of activity is very different from the scientific organisation of work theories such as were advanced by the American engineer F.W. Taylor at the turn of the century. The timing of a piece of work (for instance the assembly of a component on a machine) has now been replaced by looking at the total time spent by a team on a particular order, which involves the co-ordination of a number of activities: adjusting the equipment, stocking up with raw materials, quality control, analysis of problems, co-ordination. This indicator, used by the operators themselves, is therefore an approach which stimulates self-evaluation. So evaluation has developed until it has become an agreement that an organisation makes with itself. To ensure reliability it may be guided by external experts, but the outcome is a specific model, continuously improved, which meets its own needs.

## Figure 8.4: A Range of Performance Indicators

---

### 1. Non-financial Indicators

**Product innovation**

- Number of new product ideas, product-enhancement ideas evaluated last year

- Percentage sales/profits from products introduced in the last 3 years

- Percentage sales/profits from products with significant enhancements in the last 3 years

---

**Product development**

- Time to market
  *Average concept-to-launch time, time for each phase*
  *Average overrun, percentage of products overrunning planned finish date*
  *Average time between product enhancements, redesigns*

- Product performance
  *Product cost, technical performance, quality, return on sales, market share*

- Design performance
  *Manufacturing cost, manufacturability, testability*

**Process innovation**

- Process parameters, cost, quality, work-in-progress (WIP) levels, lead time etc.
  *Performance versus competitors*
  *Percentage improvement over 1 or 3 years*

- Installation lead times

- Number of new processes, significant process enhancements in year

- Continuous improvement
  *Number of improvement suggestions per employee, percentage implemented*
  *Average annual improvement in process parameters (quality, cost, lead time, WIP, reliability, downtime, capability)*

- Progress to lean production WIP, lead times, quality

**Technology acquisition**

- Number of licences in/out and number of patents over last three years

- Percentage of R&D projects that led to successful new/enhanced products/processes

- R&D/technology acquisition cost per new product
- Failed projects
- Percentage of projects killed off too late (after substantial expenditure)
- Cost/benefit performance of completed R&D projects

**Commitment**

- Percentage of employees aware of company innovation policies and sharing values

**2. Financial Indicators**

- Overall performance (including market share, growth rate, change in productivity)
- State of balance sheet, profit and loss, key ratios, cashflow, return on investment
- Internal and external image of the business
- Increase in employment, salary levels and increases, qualifications.

### 8.3 Financial Evaluation

This area of evaluation has already appeared at the point at which the preliminary studies were carried out, when the Consultation Group estimated the likely return on the various potential investment options. It also carried out a search to ensure that the required finance would be available from specialist institutions.

#### *Calculating Return on Equity*

An investment which is to bring a return obviously has to be able to bring in more resources than it consumes. But the business can only invest if by doing so it will not put its own finances in difficulty. That is why any evaluation needs to look at general movements of finance and not only at the expenditure and income that is due to the new equipment. The accountants need to calculate these at different points in the life cycle of the equipment: at acquisition; during its use; and when it has finally been amortised.

In order to enable comparable calculations to be carried out for equipment with varying life expectancies, the flows need to be converted to current value. The choice of discount rate is critical in this operation, since it will have a decisive impact on the result.

*Selecting a discount rate* begins with the fixing of a minimal rate of return corresponding to that offered by a risk-free investment in the financial markets. The rate used can vary from sector to sector: biochemicals, for example, in which development is rapid and risk-high, needs to work from a higher rate of return than mechanical engineering, which is a much more mature and predictable industry. This is usually an average rate. The level can be of 5 per cent or 6 per cent, if one comes up with a range from 3 per cent to 9 per cent; for a range from 6 per cent to 12 per cent, it is of 8 per cent or 9 per cent. Too high a level will discourage the taking of even small risks; whereas too low a level merely increases the risks. What is required therefore is a reasonable chance of being able to make the most of the funds available to the business given the current state of the market. This use of an average rate will give the green light to projects which don't necessarily have the highest financial rate of return, but which will create cashflow which can be recirculated even before the machinery has been fully amortised. This will therefore enable the business to continue to develop. The object is thus consistently to generate cash flow, rather than achieve exceptional results.

Three indicators are used:

**1. The Payback Period.** The payback period corresponds to the time it takes to recover the initial expenditure. If this period is short (between 18 months and 3 years for investments in productive equipment), the business is obviously able to generate the cash flow to be able to make the most of other opportunities which may arise. The calculation is thus a simple one, its object being to be able to eliminate, without further analysis, any project only likely to generate average results; or it can be used to make the choice between different projects a more consistent one.

The length of the payback period is obtained by adding together the annual incomes less the initial investment. The date at which this figure becomes positive is that at which the initial expenditure has been fully recovered.

This is, however, far from being sufficient. First, it ignores the profitability of the equipment when the time during which it is used extends beyond the recovery period. Second, the comparison between competing investments is only valid for projects of the same length, having similar total costs and looking to achieve the same objective.

So it is fair to say that the payback period measures the liquidity of the investment rather than its return. That it is so frequently used is very much a reflection of the uncertainty in the overall environment and is further proof of a high aversion to the taking of risks. Decision-makers prefer to consider any return subsequent to the date of the initial investment being recovered as very much a bonus, and if nothing is gained then this will not be seen as a major risk for the business.

**2. Net Present Value (NPV).** The net present value of a project is based on calculating, for each year since the launch, the balance of income and expenditure and discounting this to allow for the timing effect (current revenues being more valuable than future ones). If the NPV is negative, the investment will not generate a return that will cover the cost of financing it (usually taken as the discount rate). If several projects are being compared against each other, the one which achieves the highest NPV is considered to be the one with the highest rate of return.

It is not certain, however, that the project with the highest NPV will automatically be selected. A project with a lower return may be preferable, where it involves a lower risk.

Suppose that Project A has a 10 per cent chance of resulting in a negative return (-£50,000), and a 15 per cent chance of achieving the maximum return (£150,000), and a 75 per cent chance of achieving the most likely outcome (£60,000). This project is in competition with Project B, which has the following returns: 15 per cent chance of a negative return (-£100,000), 10 per cent chance of the maximum return (£180,000), and a 75 per cent chance of the likeliest outcome (£95,000). In calculating the balance of income and expenditure for each of the projects, one should initially weigh the returns for their possibility of occurrence. On this basis, Project B gives the higher return and so would be chosen. However, Project B gives rise to a worst-case scenario of a loss of £100,000. This might sink the business if it occurred (15% probability) so the company might choose Project A with its lower return on grounds of prudence.

It is worth underlining that if the NPV indicator has been calculated by the top managers, it is the production team and better still the sales team who should calculate the likelihood of each of the hypotheses occurring. Their experience is decisive and is not based on any mathematical calculation. It is in fact an expert valuation.

**Figure 8.5: Calculating Net Present Value (NPV)**

---

The most straightforward way of determining whether a project yields a return in excess of the alternative equal risk investment in traded securities is to calculate the net present value.

In formula terms the NPV can be expressed as:

$$\frac{FV^1}{1+K} + \frac{FV^2}{(1+K)^2} + \frac{FV^3}{(1+K)^3} + ... \frac{FV^n}{(1+K)^n} - IO$$

This is where IO represents the investment outlay and FV represents the future values received in years 1 to n. The rate of return (i.e. K) used is the return available on an equivalent risk security in the financial market.

---

**3. The Internal Rate of Return (IRR).** The internal rate of return of an investment project is the discounted rate for which the net present value is nil. The formula for the calculation is thus the same as for the preceding one. Only the procedure for making the calculation changes since, in this case, the unknown factor is the discounted rate and not the NPV.

This rate of return on the project needs to be compared with the rates of return which other available placement opportunities would attract. If it is greater than them then the investment project can be considered. By comparing the rates of return on several projects, it is possible to select the one with the highest rate of return. However, the decision cannot be based on this indicator alone. Despite its apparent simplicity, it contains pitfalls in terms of mathematical interpretation, since it conceals the eventual trend of the NPV which can either be rising (and therefore favourable) or falling.

The main benefit of the internal rate of return is that it is a good indicator of short-term financial viability. But it is important to keep financial viability in proportion: investment decisions also need to be taken in terms of long-term strategic objectives, which should be of greater significance. Between investments, financial viability alone does not enable a choice to be made; the overall strategy of the business needs to be the decisive argument. These indicators are a tool in choosing a solution when faced with a number of technical variations, but it is fair to say that they are only infrequently used at the outset of the strategic decision-making process.

### Looking for Finance

Understandably enough, a financier does not have the same attitude to investing in a business as does an entrepreneur. The two are in very different situations, if only because of the relative weight of their commitment. For a small company, a request for a £100,000 loan is enormous; for a banker it is probably just another Wednesday afternoon. Thus the owner of the business is primarily motivated by the hope that the innovation will succeed, while the banker starts by looking at the risk involved in the project, and the ways in which mortgages and securities can provide cover.

This fundamental difference of points of view does not mean that both parties do not need the other, and that some kind of arrangement cannot be reached. As well as the aspects referred to above, the banker will pay particular attention to three key factors: the people whose project it is; the credibility of the project itself; and the financial resources and track record which the business can demonstrate.

**The people in charge of the project**. In the eyes of the financier who is involved in lending risk capital, the experience and character of the person who has responsibility for the project, whatever their position in the company, are major factors in their ability to deliver the project. There will therefore be considerable attention paid to this person's career and probably also to the CV. Any document which is produced with a view to obtaining funds needs to contain details of the team which will see the project through, together with references.

**The credibility of the project**. Is the project going to be able to pay back what it borrows? In the short term, this is obviously all about the need to get a rapid return on any investment. No financier will make a commitment unless convinced of the commercial merits of the product compared with its current — and potential — competition; or of the fit between this particular investment and the medium-term development plans of the business. The business will need to put forward an argument built upon rigorous market research, and to justify the industrial and commercial decisions involved.

From the financial point of view, negotiation will most likely revolve around the financial plan, which takes into account the resources of the business itself, its current and (potential for additional) commitment, and its ability to repay any borrowings. There certainly needs to be a margin for error. A business plan must therefore ensure that it demonstrates the source of all the financial resources which will be needed to meet cashflow requirements.

**Figure 8.6: The Investor's Checklist**

---

- Is the company using innovation appropriately to support its business strategy?
  *What is the role of innovation in the strategy?*
  *Are the plans for innovation consistent with the company's overall strategy, product and market strategies, and other investment plans?*

- Are the company's plans for innovation workable?
  *Is the management team up to it?*
  *Does the company have the necessary human resources and skills?*
  *How strong is the company's track record?*
  *Have the plans for major projects been thought through?*
  *Is the innovation effort balanced?*
  *Is the company planning to invest a realistic amount?*

- Have risks been realistically assessed?

- What will the net financial impact be and is it accounted for appropriately?

- Do the risks and returns translate into long-term shareholder value?
  *Has the company estimated the impact on earnings?*
  *How valuable are the growth opportunities offered?*
  *How much should returns be discounted because of risk?*

---

**The financial resources of the firm and its return on capital.** The availability of cash in any business is dependent upon the demands and response of its partners, suppliers and customers. A financier will fix the first repayment date for any loan based on the payment terms which the business has both with its suppliers and its customers. This may well lead to the financier contacting a number of these partners.

If the business is a sub-contractor, a financier is unlikely to make any commitment unless the provider of the order to sub-contract can give certain guarantees: the arrangement must already have been tested, and the results in practice must tally with those from the initial study. Similarly the co-ordination in the area of computer-assisted production management will also need to be subjected to particular scrutiny; and the provider of the order will need to have made a commitment for a certain period.

To sum up, it cannot be repeated too often that the evaluation process is both necessary and never-ending; that it concerns the workforce on the shop floor as well as the managers at their desks; that it needs to take place before, during and after the implementation of an innovative project; that it involves a mixture of qualitative and quantitative factors; and that it should not be restricted to direct costs and effects.

Evaluation is thus the result of an extensive, overall understanding of the business and the project, and of the place of both in the competitive environment.

*Chapter 9*

# EVALUATING THE IMPACT OF THE PROJECT

---

*Contents*

---

*This chapter concentrates on how the evaluation of the results of a given innovation process should be carried out. This evaluation concerns therefore the overall result of activity, from a financial and marketing point of view. It is thus carried out after the event.*

*The previous chapter, like the discussion on the Diagnosis in Chapter 3, underlined the importance of adopting a methodical and critical approach to the expected results. The same approach needs to be applied throughout the process and at its conclusion.*

*But this can only take place where the information required is openly available. The key requirements for the success of this stage of the innovation process are thus transparency, critical judgement, discussion, flexibility and listening as the basis for developing a consensus which enables the organisation to move forward actively. For this to occur, the widespread tradition of limiting access to information to anyone who is not near the top of the pyramid needs to be turned on its head. A collective reflection is even more important in that nobody should be under any illusion as to the current limits of the knowledge that is shared, nor as to the absolute imperative that what the business already knows and has developed should be secured and protected (see Chapter 15).*

*The first task needs to be to gain control of the objectives that were set (and which, naturally enough, evolved in the course of the process) and of the resources which were directed towards achieving them. These will make it possible to select the impact indicators. The next step should be to use these indicators in the context of the strategic guidelines laid down both for product and process strategy, which can in turn further be divided into clear strategic paths. The impact evaluation will lead finally to a series of discussions and subsequently to conclusions which will form a new element in the collective experience of the business. The majority of the questions asked are naturally an extension of those which were raised in the previous chapter. It is for this reason that this chapter is a short one, and concentrates on general issues rather than on specific techniques.*

## 9.1 Why Evaluation Matters

Experience shows that in the majority of cases it is the ***evaluation stage of an innovation which is most frequently skated over, cut short or even totally ignored***. With the possible exception of financial issues like turnover and profit, there just isn't enough time for it. Evaluation, to many managers, looks too much like a form of academic speculation which is intended either to make one feel good (in the case of success) or to give the opportunity for a bit of self-flagellation (in the case of failure). The frequently stated view is that there is no time in business for this kind of exercise. From the point at which the investment has been made (and can no longer be unmade), the key issue becomes one of making the best out of what one has got, and so any evaluation of its impact sometimes appears to be a side-show to real life.

There are certainly exceptions to this attitude, but they tend to be extreme cases: the real success stories or the total disasters. In the first case, reality has exceeded all expectations, and the questions from journalists, the authorities, neighbours, partners and employees require some kind of an answer. These answers call for reflection, and thus for evaluation. It is often forgotten that these responses inevitably distort reality. Journalists naturally latch on to the details which make a good story. Financial or technological success creates an aura of achievement, and it is natural enough for everyone involved to embellish their own role in the success story.

In the total disaster situation, on the other hand, and in particular where individual jobs are threatened, or where the business has radically to be restructured if it is not already bankrupt, each type of partner — whether employees, bankers or shareholders — is asking what went wrong. In this case, analysis is invariably the basis of any defence.

These two extreme cases are, however, relatively rare and are perhaps more likely to result in a justification of what happened than in an in-depth and frequently contradictory evaluation, leading to a comprehensive list of the "pros" and the "cons". In the case of 90 per cent of innovations, there is no pressure from outside to carry out an evaluation, and the process tends to tail off before reaching firm conclusions. New projects and new pressures mean that "there is no time to evaluate". This is, however, a major mistake.

**Figure 9.1: Examples of Different Levels of Evaluation**

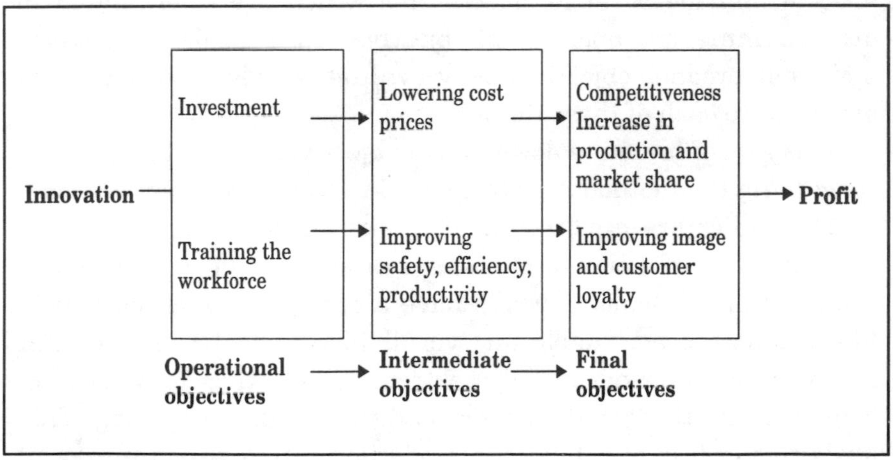

## 9.2 The Target and the Indicator

The indicators against which a project is evaluated must be appropriate to the intended objective. For example, a company which buys a robot to do all the packaging may have done so with the objective of satisfying a key customer, such as a supermarket chain. But it was aiming, at the same time, to:

- Increase the speed of deliveries

- Reduce the level of damaged returns

- Reduce the employee overhead

- Eliminate one or more bottlenecks.

In the majority of cases, the initial objectives are able to have a positive impact on the other objectives, enabling several birds to be killed with a single stone. In this particular example the introduction of a new piece of equipment will have initially enabled the ideas and demands of the customer to be satisfied, but it will rapidly have led to changes elsewhere in the organisation. The target is thus complex and always on the move; and this is why the final target for evaluation is not necessarily the innovation itself, which is simply a means to an end. The effect of an infatuation with everything that is "high tech" tends to place too much attention on the innovation itself and thus to neglect the overall impact which it has produced downstream. It is the role of a manager to manage a range of activities which are co-ordinated to achieve immediate operational objectives, intermediate objectives and final broader objectives. Any evaluation needs to pay equal attention to each of these three areas.

This initial breaking down of the objectives raises three issues concerning the methodology to be used in this form of evaluation.

The indicators are first of all instruments to carry out the evaluation of an action or of a series of actions. They represent a combination of elements which have a common value and which, like a macroscope, enable an overall picture to be viewed. But these same indicators are also extremely selective: they may reveal only one thing, but they pretend to explain everything. This leads to the loss of a huge mass of information which may well include potentially crucial data.

This is the reason why it is important to remember that indicators can distort reality and contain the ever-present risk of providing only a partial diagnosis or even an incorrect verdict.

A qualitative or quantitative indicator is always built upon the shifting sand of subjective choices.

A second problem applies specifically to the quantitative indicators. These are the most often used, but they always represent fairly insignificant average values. Two businesses with the same average productivity may, nevertheless, be in very different states, and have to adopt very different strategies if they are to improve their position, depending on whether productivity levels are consistent or whether they have a few highly productive individuals whose value is concealed by considerably less productive colleagues. A good indicator is not one which merely gives the average or the median figure, but also highlights the range which these figures conceal.

The third and final problem is to do with qualitative indicators, which tend to be even more difficult to pin down than the quantitative. There is frequently no real sense of where one ends and the next begins; it is more of a continuum since qualitative indicators need to be grouped together, so that there is some relationship between the findings. A qualitative indicator on an innovation issue can be obtained from putting together a questionnaire which takes into account the views of the workforce, their managers and their customers on the results obtained. This indicator can be supplemented by quantitative indicators. The difference between the two types of data can in turn be used as a measurement of the subjective opinion which the employees have of the innovation in question, and thus provide useful leads as to how the organisational problems which the innovation has caused can be identified, defined and resolved.

## 9.3 A Few Impact Indicators

There is thus no all-purpose set of indicators suitable for use in all places and in all weathers. Each business, and each innovation within it, is a special case. Any indicators of the impact of a project need to be established for a specific situation, and to fit the strategic objectives which will themselves, it is hoped, have been defined and communicated at the outset (in particular by the leader of the Consultation Group).

With a view to providing a few examples, this section provides an overview of indicators which should be a useful guide in the case of particular strategies being pursued. These are divided using the classic (but nevertheless arbitrary) distinctions between product strategies and process strategies, and between offensive strategies and defensive strategies.

### Product Strategies

Product innovations can be of a defensive or an offensive nature.

Where they are *defensive,* they often emerge out of a loss of customers or of market share due to a particular obsolescence in a product range (see Chapter 10 on the concept of obsolescence). The number of customers may have been in decline, and product recognition may also have fallen; the indicator must be seen as the signal that some kind of defensive action needs to be undertaken against increasingly predatory competition. The first indicator to take into account is the number of customers (market share often being a difficult figure to calculate for a small-medium company, except where the business is active in an easily defined and measured niche), making the comparison between the state of affairs before action is taken and the change which follows the introduction of the innovation. To this measurement could be added an estimation of the volume of the new product sold compared to the old, and a number of purely financial indicators: turnover of the new product as a percentage of total turnover; turnover of the new product in relation to the cost of the innovation process. The key problem that needs always to be resolved in this situation is the influence on the change which has occurred of external factors (such as the overall state of the economy, or changes in buyer behaviour that have affected the market as a whole).

The objective of any *offensive* measures is to anticipate competitor activity. In this case, it is a question of responding to market demand in advance (or maybe even creating market demand), by looking to gain some control over the uncertain. For a business which is an "early bird" (see the box below), moving forward by experiment, or exploring new trends, may often be their best means of survival. In this case, the best indicator could be the new product as a percentage of total sales, or the number of new customers which it attracts. But at the same time it is important that there are also qualitative indicators, such as the "value" of these new customers. These indicators also need to be set against financial indicators which have already been defined.

## Figure 9.2: Company Characteristics Based on Market Performance

A key element in impact evaluation is being able to make comparisons with other companies. Doyle, Saunders and Wong (1986) define six types of company sharing particular strategic and organisational characteristics:

- **Early birds.** Highly successful. These are among the first companies to adopt new technologies and the earliest to enter markets. They enter aggressively, focusing on expanding primary demand and opening up new segments. Strong on marketing, they seek clearly-differentiated quality positioning in the market. They spend more on promotion and advertising. Prices are at competitive levels or higher. Highly-committed managers with clear marketing strategies.

- **Price Fighters.** Successful. Aggressive pursuit of market share. Strong marketing skills and clear strategy of being price leader with modern product range. Products designed for low-cost manufacture.

- **Cruisers.** Good performance. Mid-positioned between Early Birds and Price Fighters with strategy based upon broad range of quality products at competitive prices. Slower to change direction, but highly-efficient in manufacturing and marketing.

- **Sprinters.** Moderately successful. Less innovative than the Early Birds, but fast followers. Once a market is proved, they rely upon aggressive marketing and flexible manufacturing skills to catch up. Risk-averse and profit-conscious.

- **Mastercraftsmen.** Poor performance. Traditional produc-tion-oriented companies. Product quality is high but pro-fessional marketing non-existent. They lack customer knowledge, competitive positioning and designing for the market. Least aggressive of all the groups. Low promotional spending. Tightly-defined jobs.

- **Lemmings.** Bad performance. Older, declining companies. Lacking marketing skills and product quality. Sometimes pulled into new markets, but without marketing, techno-logical and innovative abilities they make little progress. Little value added.

In contrasting the impact of successful and unsuccessful groups, four key differences emerge:

- **Professional marketing.** Good products need to be allied to professional marketing skills to exploit them through forceful market segmentation and positioning strategies.

- **Decisive entry strategies.** Successful companies enter markets or technologies earlier because they identify clear opportunities.

- **Commitment to market share.** All the successes have long-term strategies, rather than over-focusing on short-term profits.

- **Organisational commitment.** A belief in group involve-ment in formulating strategy, and continuous monitoring of progress in the market.

### Process Strategies

In the case of process strategies, we once again come across the same distinction between defensive and offensive strategies.

*Defensive* strategies tend to start from the observation that there is a stagnation (in a growing market) or a decline in turn-over for a range of products which are far from being obsolete. The product can still perhaps be termed a success, but is being displaced by competitors who are undercutting on price, meeting orders more rapidly or adapting with greater ease to changes in the patterns of demand.

*Offensive strategies*: the objective of any innovation is to increase profits by exploiting the relative weakness of the competition. This objective may be achieved by a strategy aimed at a rapid increase in profits (which may be the key quantitative indicator), or it may be decided to develop a strategy which begins by aiming to increase market share or to eliminate particular competitors (in which case the key indicators become the number of customers or the turnover taken away from a targeted group of competitors). In this case, it is clear that the product is developing in the right direction, and that the innovation can be implemented rapidly, at a lower price, or in a more flexible way. Once the financial calculations have been made, the company knows whether it will be successful, and also when.

However simplistic it may be, this classification of strategies can make the selection of indicators an easier process. But, once again, it needs to be said that none of these indicators contains the absolute truth. The figure below contains a summary of the suggestions contained in this chapter; but needs to be read with the health warning that any proof which it offers as to the right decision having been made needs to be weighed with the equally significant evidence provided by the financial evaluation which has already taken place (see Figure 9.3).

## 9.4 How to Make the Most of Indicators and Evaluation

There are three universal truths which must always be born in mind:

- No quantitative indicator is perfect, but this is never a reason for ignoring them.

- General indicators are very useful but should be replaced by those which focus on the impact of the innovation in question.

- Evaluation is a process and so it can therefore be primarily used to modify or revitalise behaviour.

**Figure 9.3: A Synthesis of the Different Elements of Impact Evaluation and Their Indicators**

|  | **Impacts** | **Indicators** |
|---|---|---|
| **Process strategy** (Defensive/offensive) | Increase in profits | • Margin<br>• Profit ratio |
|  | Increase in market share | • Turnover<br>• Market share/product |
|  | Elimination of competition | • Number of customers<br>• New turnover/turnover |
| **Product strategy** (Defensive/offensive) | New markets | • Marketing indicators<br>• Number of customers<br>• Market share<br>• Trends<br>• Market share of new product<br>• New turnover/total turnover |
|  | Existing markets | • New turnover/cost of innovation<br>• Share of sales new product<br>• Number of new customers<br>• Value of new customers |
| **Internal social relations** | Increase in qualifications | • Hours lost to strike<br>• Level of satisfaction at all levels<br>• Training |
|  | Size of labour force | • Absenteeism, staff turnover |
| **Involvement in the community** | Spin-offs | • Relations with competitors<br>• Relations with families of staff<br>• Relations with administrations |
| **Productivity** | Sale of technology | • Value added/employee<br>• Potential value added/employee |
|  | R&D | • Hidden conflicts<br>• Overall productivity indicators |

Evaluation, in the form that it has been described here, concentrates on innovation which has a precise object within a clear time frame. The expenditure which it involves is clear and the results obtained both identifiable and measurable. Unfortunately, however, it is not possible to be blindly confident in these impact indicators. Innovation by its nature invariably has an impact on too many aspects of the life of a business: internal and social tensions, productivity, the participation and satisfaction of management and workforce, the shape of the organisation and the balance of power within it, and the cohesion of the whole. It also changes the impact of the external environment, and relationships with the customer base, both in quantitative terms and in terms of satisfaction.

This broad range of factors needs to be taken into account if an evaluation is to be made of the long-term effectiveness of a chosen strategy. Hence the importance which needs to be given to the creative (as opposed to procedural) character of the selection of the right indicators. In this context, "procedural" is understood as something that can be reduced to a mechanical list of operations to be carried out.

Measuring the extent of internal tension is not simply a question of taking a stopwatch and measuring absenteeism, for instance, which may be due to a range of seasonal and other external factors. It needs to be extended to consider the satisfaction of the workforce and the openness to change of employees. Similarly, to evaluate social conflicts requires considerable tact and skill, while to evaluate productivity is far more than a crude calculation of turnover per employee, but also needs to take into account the potential turnover per employee. All of which leads to the larger issues of climate and commitment.

The medium-term social impact of any strategic decision is thus an excellent barometer. Its analysis may well reveal inconsistencies in management or in the chosen objectives in areas where the decision itself appeared entirely logical; or it may throw light on indicators other than those which initially appeared to be the most revealing. At the same time, it should not be forgotten that the exclusive, if multi-faceted, evaluation of the effects of an innovation process can probably never reveal the diversity of the results (both positive and negative) in the same way as can an evaluation of both cause and effect which also takes into account the timescale and the different stages through which the project passed.

# Chapter 10

# ANTICIPATING OBSOLESCENCE

*Contents*

*There was a gap of 81 years between Lebon's invention of the first piston engine and the production of the first four stroke engine by Daimler.*

*There was a gap of 28 years between the first steam locomotive and Stephenson's Rocket.*

*There was a gap of 17 years between the invention of the turbo-jet and the breaking of the sound barrier.*

*Today, one pharmaceutical product in two did not even exist five years ago.*

*The accelerating pace of change is clear enough, but things sometimes move even more rapidly. The most powerful super-computer in the world, when it was installed by Thinking Machines in Chicago, only held on to its record for 15 days before an even more powerful computer, produced by a competitor, elbowed it aside.*

*Nothing is new for long in a world where speed of change is all. This is the price that we pay for open markets, for market economies, for technology. Technology is increasingly at the centre of this progress, and increasingly rapidly pushed out of the spotlight by other technologies, which themselves are replaced at even faster speed. Innovation breeds innovation. The accelerating pace of technological change is a complex issue, but this is not the place to describe the processes of its creation, of its development on an industrial scale and of its distribution. Concentrating on the business itself, this chapter will examine it in terms of competitiveness and the change process. On one side, competitors are more and more aggressive in their use of innovation to achieve a direct impact on the marketplace; on the other, employees themselves participate more and more in the development of their organisation. Innovation has an impact on the whole business, and in turn the whole of the business accelerates the need for innovation.*

*There is a further paradox: the more innovative a technology, the more rapidly it ages, and thus the more rapidly it moves through its life cycle.*

*It is for this reason that it is essential to know how to manage the maturity and eventual decline of a process or of a product; and particularly where this may have been, for a time, the spearhead of the business.*

*It is also essential to be able to identify all the "secondary obsolescence" with which any business is riddled.*

*This chapter begins by defining the concept, before moving on to examine the forms of obsolescence which afflict the normal working of a business, and which are here called "easy" obsolescence. The discussion then turns to strategic obsolescence, which is to do with the relations between the business and its external partners, and which is here termed "difficult" obsolescence.*

## 10.1 Every Innovation Carries the Seeds of its Own Obsolescence

Obsolescence literally means something which is no longer used. But in any business, how many of these obsolete objects, components or practices are still in daily use? The issue needs to be examined on different levels. On the one hand, there is obsolescence in objects; and on the other, an obsolescence in organisations and human relations. There is product obsolescence, and there is obsolescence in market relationships. Finally there is strategic obsolescence, where it is long-term behaviour that is the cause.

It is not difficult to recognise obsolete objects: the life cycle of a machine is usually known, and it is not difficult to keep up-to-date on new production processes, or to put together a flexible depreciation plan. Nor is it a particularly difficult task to identify obsolescence in a particular product compared to its market, either through using quantitative indicators, or through personal experience and awareness.

It is much harder, however, for psychological and emotional reasons, rapidly to notice the obsolescence of a whole product range or of a whole sector around which the entire business may well be based. This is still more true for obsolescence in human relations and personal relationships within the business.

In its everyday use, the term obsolescence has a negative connotation. An obsolete piece of machinery is viewed as a useless antique (*ferrovecchio* in Italian). But if an object is obsolete today, it is because it once had a value; and if it is ill-adapted or inefficient today, it was once probably suitable and efficient. A negative perception now leads too often to a positive judgement of the past as a whole, without really bothering to search for and examine the deeper causes.

It is, however, well worth making the effort, and will certainly be useful for what follows, to consider briefly the distinction between functional internal obsolescence and a more general strategic obsolescence. The former consists mainly of a loss of efficiency in a specific piece of equipment or in a production system, within an overall process which continues to be dynamic. The latter concerns the total system which is the business and is effectively the combination of a variety of elements. It may be less dramatic and less visible, but it is infinitely more dangerous for the life expectancy of the whole edifice.

Innovation and obsolescence are closely related. The first is the cause of the second; its effect is both to plan for and to anticipate obsolescence.

Unlike the term obsolescence, that of innovation has a positive feel to it. But although innovation is considered to be positive, this is not always the case. It is not the case when a brilliant innovation results in a financial disaster (like Concorde) or, on a smaller scale, when it is less suitable than other more traditional technologies used in an original way, and leads a small business to financial ruin. Nor is innovation necessarily viewed with enthusiasm at a national or regional level when it results in a smaller workforce and a larger unemployment figure. In this case the competitive environment highlights a conflict of interest between the business and the national position. Lastly, in everyday language "new" is normally used as a synonym for "undeniably good"; hence the frequency with which it appears in the language of politics.

Ideally, a perfect synchronisation should exist between the staged introduction of innovation and the control of its obsolescence. However, even in situations where there are excellent indicators to assist with the choice of innovation and a programme of maintenance to monitor obsolescence, most businesses tend to operate what can most charitably be described as a "flying blind" policy. The use of indicators was discussed above in Chapters 8 and 9, and the point was made that, when selected in terms of specific objectives, they can be used as management tools within a key idea of *fighting obsolescence*, which will require the detailed development of a maintenance policy. But what should be done when faced with unstable or extraordinary conditions in which the external environment changes abruptly and unpredictably?

## 10.2 "Easy" Obsolescence

Internal obsolescence can be planned for using the life cycle of machines and their maintenance requirements; but strategic choices also have a role to play in this process. With this aim in mind, the whole system of indicators, which has already been described elsewhere, can also be used as the means of monitoring obsolescence.

*Maintenance* is a term in daily use in a number of sectors, but rarely uttered in others. It is mainly used with reference to keeping technical objects, and in particular machines, in good working order. Maintenance is increasingly a task demanding serious qualifications, involving repair jobs and a maintenance function. While the traditional maintenance worker continues to arrive in a boiler suit clutching an oil can, the electronic maintenance worker wears a suit and removes dusting instruments and a screwdriver from her briefcase. But the aims are the same: to maintain the equipment before it breaks down so as to ensure that it maximises output and use of time. Our civilisation has shifted away from being characterised by physical tiredness to one of mechanical breakdown and electronic failure.

It is important to underline the different rhythms of change. As a general rule, new processes and technologies bring about a high rate of change, whereas mature technologies and processes change at a slower rate. But this is also a problem that concerns products: even if a product is made up of already mature technologies, it can very rapidly become the object of redefinition (of style, design, option, function). It is thus clear that behaviour with regard to sectors having a different speed of development (speed of product or of process, or both) imposes different "philosophies", which goes well beyond the question of skills levels.

There is another type of maintenance which creates quite different problems: ***maintenance of organisational systems***. The systems within a business, like those within a town, are organised by general or specific "projects" structured around defined objectives and linked into an overall process. The idea of maintenance is too often considered to be an extraordinary act, and is associated with a very different and even directly opposed act, namely repair.

To resume the argument: a system such as a business needs, in normal conditions, a permanent maintenance programme, which has both a preventative and corrective role, and has an impact both at the level of its machines and of its organisation.

This control system cannot wholly be put into place by the business itself. The role of the business is to "think it", but it probably needs to call on external advisory support.

It is also necessary to take account of the fact that, in certain sectors, the development of technology takes place at such a speed (for instance in the field of microelectronics) that it is practically impossible to follow the rhythm of technical change and that it is much easier to be pushed along at a ridiculous rate, which is imposed by producers and by fashion, than it is worth taking carefully considered investment decisions. Here again can be seen the usefulness of specialist, disinterested agencies which are able to assess and evaluate rapidly changing processes — in short, to benchmark them (see the box below). They also enable the forging of completely new (if one knows how to make the most of them) partnership opportunities.

## Figure 10.1: The Benchmarking Process

Benchmarking is not simply making a visit to another company to see what they do. Like innovation, it is a structured, analytical, continuous process which seeks out the best practices that will lead to improved performance. There are a number of reasons why benchmarking should be used:

- Integrating the best ideas from several firms will result in "better than the best"

- Improvements will be incorporated quicker by shortening the learning curve

- The cost of improvements will be less when adopting ideas from benchmark companies

- The benchmark will help focus everyone in a company on continuous improvement

- Avoids re-inventing the wheel.

The process involves not only benchmarking competitors but also competitive best practices wherever they occur. This cannot be done by some haphazard, hit-or-miss method. It requires a clear step-by-step process to be effective. There is no shortage of benchmarking models, but here is a six-stage sequence that can have a real impact:

- **Identify benchmark elements**
  There is almost no limit to the elements in a business that can be benchmarked. The common factor is that the elements must be measurable.

- **Select key measurements**
  For each element identified above, it will be necessary to select one or more indicator and unit of measurement. In some cases, these measurements will have to be created because they do not exist in the business at present.

- **Measure current performance**
  Before visiting the benchmark company it is best to measure current performance. This step will reveal much about current practices and assist with preparation for the benchmark visit. More focused questions can be prepared to get a deeper understanding of how the benchmark company has achieved their level of excellence.

- **Decide benchmark partners**
  It is important to develop a partnership approach to benchmarking. There are a number of ways, described throughout this book, of discovering which companies have the best practices. Benchmarking is not necessarily confined to a similar industry.

- **Visit and collect data**
  Once the preparation has been completed, the benchmark company can be visited and data collected on the elements selected. It is important to discover not only *what* results have been achieved at the benchmark company, but also *how* they have achieved them.

- **Analyse results and implement improvements**
  The mass of data collected needs to be analysed and turned into useful information. The need to ensure true comparisons (carrots for carrots) is essential, together with an understanding of cause and effect. Action plans can then be established and changes implemented to improve performance.

When all these steps have been completed and improvements implemented, it will be necessary to track progress against benchmark elements. Benchmarking is a never-ending process because other companies continue to improve over time.

Finally, it is important to underline, in what is very much a bridge between "easy" obsolescence and "difficult" obsolescence, the question of *maintaining organisations*. Any social structure, even the most perfect and the most efficient, is exposed to continuous and extremely powerful restraining forces. The conflict between individual and collective short-term interests is never-ending. Even the best thought-out and constructed organisational charts can do little more than define responsibilities and control mechanisms, and provide a kind of justification for individual or group situations. They may enable changes in behaviour to occur, but only for a time. Very quickly, routine comes rushing back. Very quickly, the imposed structure transforms a factor for change into a force for inertia.

A recent example illustrates this argument: during the great wave of nationalisations which took place in France in 1982, a large number of the managers in those organisations likely to be nationalised (who had in the main voted against the newly elected left-wing government), changed their behaviour and considered the nationalisations to be their new route to security. They finally had the feeling that things were moving, and that they were being looked after. The change of structure was in effect a real clear-out. It began with the naming of the new managing directors, followed by that of the senior managers and by a series of behavioural changes which cascaded down to affect all levels of the workforce. It led to a fundamental change of attitude, which had a highly salutary effect on the way that some of these industrial dinosaurs evolved. But astonishingly, when a number of these same businesses were to be denationalised a few years later (in the period beginning in 1987-88) a similar sense of energy and dynamic purpose was again, paradoxically, generated. The same managers threw themselves with equal energy into the denationalisation process as they had into that of nationalisation. This is not a judgement on nationalisation, but merely highlights the positive effects which accompany organisational change.

## 10.3 "Difficult" Obsolescence

We now turn to general or strategic obsolescence, which leads one to look further and to enter the arena of strategic behaviour in the broadest sense. To maintain an open system one must not limit one's ambition simply to maintaining the existing one. It is not sufficient to make corrections in the present in the midst of rapid changes in the areas of demands, tastes and technologies. It is necessary to integrate an ordinary maintenance programme with strategic maintenance, which is through this process related to the future.

Maintenance of a business system over time requires three conditions to be fulfilled:

- Direct involvement of the workforce, who consider it a creative and fulfilling duty to defend the overall state of health of the business

- A balanced and mutually beneficial relationship with customers who are equally interested in nourishing this relationship

- Availability of an efficient information system, with a network of indicators linked to a management plan which provides real-time information on the external environment and on the internal state of the system.

Flexibility and specialisation are two apparently contradictory methods to avoid the fatal consequences of the type which hit the sail-makers in the course of the tragic decline of the sailing boat industry. Flexibility consisted in being open to experimenting with gyroscopes and manometers, even when the sails were doing the work. To have been far-sighted would have been to understand that all sailing boats would be replaced by steam boats and that there still existed a few possibilities for speedy clippers. It did not consist of imagining that there would exist, in 60 years time, a new market for sailing pleasure boats.

So strategic obsolescence is a question of turning one's gaze away from the everyday, and taking a creative view of a much broader vision. Action can most usefully be complemented by investment in research activities.

Michel Piore and Charles Sabel (1984) have demonstrated that there exist three ways through which a business can secure its place within flexible specialisation:

1. **Participation in regional conglomerates.** The specialist industrial districts of north and central Italy, New York's rag district, and the construction industry in virtually all American cities are examples of this. In this form of association, no one business has absolute dominance, agreements take the form of contracts covering a relatively short period, and the role of the different partners can change from one contract to another. Political, ethnic and religious institutions all have a key role to play.

2. **The business federation**: for example, the pre-war Japanese zaibatsu, and the business federations set up, with a more supple structure, in post-war Japan. These associations are defined in social and economic terms. The groups are not formally integrated with financial groupings, and there is no hierarchical relationship between the different businesses. But there is a very strong sense of shared identity, and as a result of this the chances of success are greater.

3. **Associations of workshops.** In the past, the specialist steelworks of Saint Etienne or the large textile factories of Lyon operated on this model. More recent examples exist in Germany and in the United States, even if they are often difficult to identify due to their similarity with the strategies of certain major groups. These businesses do not produce standard products, and their size is often due to the size of the investment in equipment that their manufacturing process requires, rather than to economies of scale. Lastly, their suppliers are viewed as partners rather than being treated as inferiors.

# Readings and Arguments 2

The essay which concluded Part 1 of this book highlighted the difference between entrepreneurship and management. This second essay moves on to consider some recent and influential writing on the innovation process itself, and notably on the preparation, implementation and evaluation of a substantial change project.

---

### Key Arguments

- A range of potential innovations lie dormant within any business.

- Thinking about objectives and qualitative issues is essential in terms of the capacity to master new technologies or move into new markets.

- A key aspect of the Diagnosis process (itself a strategic act) is its potential to destabilise.

- The role of the Consultation Group involves a thorough examination of the general outlook for the business and of the threats and opportunities that it contains, before moving on to develop a strategic project which is built around the active strengths of the business, its long-term objectives, and its capacity to adapt.

- Any decision-making process needs to be both progressive and iterative.

- Implementation requires a strategic decision to be cascaded throughout the whole organisation, and to have a sustainable and measurable impact on behaviour.

- There is no technological innovation without organisational innovation.

- Good strategies are unambiguous, based upon detailed knowledge, realistic with regard to the capabilities of the business, and implemented with commitment.

- The innovative organisation creates the space in which individual and group change takes place.

- Which comes first, training or growth?

- The human factor is too often neglected during the implementation phase.

- A key role of management is to keep stress levels acceptable during the change process.

- Contingency planning for managing the unexpected is an essential part of implementation.

- Evaluation has to become an agreement that an organisation makes with itself, and its key partners.

- Precise evaluation of any project requires the widest circulation of accurate information.

- Every innovation carries the seeds of its own obsolescence.

We have already encountered Schumpeter's trail-blazing entrepreneur. This heroic individual has a role in innovation in general, but also in the change process in particular:

> The entrepreneur . . . may indeed be called the most rational and egotistical of all. Conscious rationality enters much more into the carrying out of new plans, which themselves have to be worked out before they can be acted upon, than into the mere running of an established business, which is largely a matter of routine. And the typical entrepreneur is more self-centred than other types, because he relies less than they do on tradition . . . and because his characteristic task consists precisely in breaking up old, and creating new, tradition (Schumpeter, 1926).

This need to play the double role of iconoclast and creator is often forgotten in the desire of the majority of leaders to be thought of as either one or the other. Yet both approaches play their part in the sequence of analysis and activity that Part 2 has described. But the iconoclast is perhaps the first to make an appearance, constantly sceptical with regard to attitudes such as "That's the way we've always done it" and "That would never work here". Levitt's famous article "Marketing Myopia" should be compulsory reading (and regular rereading) for all strategic decision-makers, whatever the size and nature of their firm or their decisions:

> Every major industry was once a growth industry. . . . In every case the reason growth is threatened, slowed, or stopped is not because the market is saturated. It is because there has been a failure of management . . .

> In each case its assumed strength lay in the apparently unchallenged superiority of its product. There appeared to be no effective substitute for it. It was itself a runaway substitute for the product it so triumphantly replaced. Yet one after another of these celebrated industries has come under a shadow.

> In truth, *there is no such thing* as a growth industry, I believe. There are only companies organised and operated to create and capitalise on growth opportunities. Industries that assume themselves to be riding some automatic growth escalator invariably descend into stagnation . . .

> The view that an industry is a customer-satisfying process, not a goods-producing process, is vital for all businessmen to understand. An industry begins with the customer and his needs, not with a patent, a raw material, or a selling skill. Given the customer's need, the industry develops backwards, first concerning itself with the *physical* delivery of customer satisfactions. Then it moves back further to *creating* the things by which these satisfactions are in part achieved. . . . Finally, the industry moves back still further to *finding* the raw materials necessary for making its products.

> The irony of some industries oriented toward technical research and development is that scientists who occupy the

high executive positions are totally unscientific when it comes to defining their companies' overall needs and purposes. They violate the first two rules of the scientific method — being aware of and defining their companies' problems, and then developing testable hypotheses about solving them. They are scientific only about the convenient things, such as laboratory and product experiments. The reason that the customer (and the satisfaction of his deepest needs) is not considered as being "the problem" is not because there is any certain belief that no such problem exists, but because an organisational lifetime has conditioned management to look in the opposite direction (Levitt, 1960).

This self-satisfaction, upon which the disinclination to respond to customer needs is based, is thus the reason why Schumpeter's entrepreneur needs constantly to find new ways — of which the Diagnosis is one — of disrupting organisational complacency. Another form which this move towards adopting the customer's point of view can assume is "process innovation", as recently defined by Davenport:

Adopting a process view of the business — a key aspect of process innovation — represents a revolutionary change in perspective: it amounts to turning the organisation on its head, or at least on its side. . . . A process is simply a structured, measured set of activities designed to produce a specified output for a particular customer or market. It implies a strong emphasis on how work is done within an organisation, in contrast to a product focus's emphasis on what.

A process is thus a specific ordering of work activities across time and place, with a beginning, an end, and clearly identified inputs and outputs: a structure for action.

Process structure can be distinguished from more hierarchical and vertical versions of structure. Whereas an organisation's hierarchical structure is typically a slice-in-time view of responsibilities and reporting relationships, its process structure is a dynamic view of how the organisation delivers value (Davenport, 1993).

And this delivery of value is not just to shareholders, but to all stakeholders, including customers:

> A process approach to business also implies a relatively heavy emphasis on improving how work is done, in contrast to a focus on which specific products or services are delivered to customers. Successful organisations must, of course, both offer quality products or services and employ effective, efficient processes for producing and selling them *(ibid.)*.

Japanese companies spend twice as much as do American companies on researching and developing new processes as opposed to new products; and almost all of this American spending on processes goes on engineering and manufacturing rather than on processes devoted to marketing and sales:

> Adopting a process perspective means creating a balance between product and process investments, with attention to work activities on and off the shop floor . . .

> Processes also need clearly defined owners to be responsible for design and execution and for ensuring that customer needs are met. The difficulty in defining ownership, of course, is that processes seldom follow existing boundaries of organisational power and authority. Process ownership must be seen as an additional or alternative dimension of the formal organisational structure that, during periods of radical process change, takes precedence over other dimensions of structure *(ibid.)*.

What this situation demands, therefore, is the ability — across the organisation — to anticipate, plan for and manage radical change, based upon a combination of what are often termed process and expert skills. Waterman tidily links the two areas of anticipation and organisation that underpin the argument and behaviour of Part 2 of this book:

> More than anything else . . . their strategy anticipates the needs of customers and innovates to keep them more than pleased. What makes their strategy . . . strategic is how well they have organised themselves to innovate (Waterman, 1994).

Waterman is not alone in the difficulty he encounters when it comes to describing strategy. Grant, for example, confidently attacks the problem. But, like many definitions, his interpretation raises more questions than it resolves:

> The task of business strategy . . . is to determine how the firm will deploy its resources within its environment, to select the organisational structure and management systems needed to effectively implement that strategy, and so satisfy its long-term goals . . . (Grant, 1991).

This is not to criticise Grant's book, which is one of the most lucid to have appeared in recent years. But it does highlight the difficulties that the majority of managers experience when asked to outline a strategy that can never be detached from, upstream, the objectives of the organisation or the environment in which it navigates and, downstream, from the shape which the organisation has assumed (or ought to assume) to be able to implement and evaluate the strategy:

> The prerequisite for a strategy to be successful is that of fit: the strategy must be consistent with the firm's goals and values, with its organisation and systems, with its resources, and with its environment . . .

> The external environment of the firm comprises all external influences which influence the firm's performance and decisions. These include the economy, the political structure, the social system, and the state of technology. However, for most business strategy decisions, the relevant part of the firm's environment is its industry and is defined by the firm's network of relationships with competitors, suppliers and customers *(ibid.)*.

Which brings us, inevitably, to Michael Porter, whose impact can be measured by the fact that his books never need to go into paperback. Whether the purchasers actually read Porter is not a question for this essay; but his influence over the last 15 years, not least through his much-misused "five forces" model is considerable.

One area of Porter's output is of particular relevance to small-medium businesses. This is his definition of "three generic strategies for achieving above-average performance". His argument contains two elements: firstly, you need to choose a strategy, or risk (like too many businesses, of all sizes) getting "stuck in the middle"; secondly, you need to choose the right strategy in order to have a chance of achieving the fit described by Grant. Porter's three strategic options are cost leadership, differentiation, and focus (with this last option having two variants, cost focus and differentiation focus):

> The notion underlying the concept of generic strategies is that competitive advantage is at the heart of any strategy, and achieving competitive advantage requires a firm to make a choice — if a firm is to attain a competitive advantage, it must make a choice about the type of competitive advantage it seeks to attain and the scope within which it will attain it. Being "all things to all people" is a recipe for strategic mediocrity and below-average performance, because it often means that a firm has no competitive advantage at all . . .

> Cost leadership is perhaps the clearest of the three generic strategies. In it, a firm sets out to become the low-cost producer in its industry. The firm has a broad scope and serves many industry segments, and may even operate in related industries — the firm's breadth is often important to its cost advantage. The sources of cost advantage are varied and depend on the structure of the industry. They may include the pursuit of economies of scale, proprietary technology, preferential access to raw materials, and other factors . . . Low-cost producer status involves more than just going down the learning curve. A low-cost producer must find and exploit all sources of cost advantage. Low-cost producers typically sell a standard, or no-frills, product and place considerable emphasis on reaping scale or absolute cost advantages from all sources . . . (Porter, 1985).

This is clearly not an approach which is likely to be the right one for the vast majority of small-medium firms. But it needs to be identified as a possible option, if only to ensure that is carefully avoided. The list of failed companies which drifted away from their original strategy towards the unrealistic pursuit of cost advantage is a substantial one.

> The second generic strategy is differentiation. In a differentiation strategy, a firm seeks to be unique in its industry along some dimensions that are widely valued by buyers. It selects one or more attributes that many buyers in an industry perceive as important, and uniquely positions itself to meet those needs. It is rewarded for its uniqueness with a premium price . . . (*ibid.*).

This is clearly a strategic approach that is more likely to work for the small-medium firm. But there are many examples of companies which failed because the price premium achieved is less than the cost of differentiating. Nevertheless, as Porter points out:

> . . . in contrast to cost leadership, however, there can be more than one successful differentiation strategy in an industry if there are a number of attributes that are widely valued by buyers (*ibid.*).

The third generic strategy is focus, with the objective being to identify, occupy and defend a niche:

> The focuser selects a segment or group of segments in the industry and tailors its strategy to serving them to the exclusion of others. By optimising its strategy for the target segments, the focuser seeks to achieve a competitive advantage in its target segments even though it does not possess a competitive advantage overall.

> The focus strategy has two variants. In cost focus a firm seeks a cost advantage in its target segment, while in differentiation focus a firm seeks differentiation in its target segment. Both variants of the focus strategy rest on differences between a focuser's target segments and other segments in the industry. The target segments must either have buyers with unusual needs or else the production and delivery system that best serves the target segment must differ from that of other industry segments (*ibid.*).

Porter is rarely anything less than thought-provoking; but Hendry and his colleagues provide a valuable synthesis of the seven modes of adaptation or development that are most appropriate for the small-medium firm:

1. Capitalising on environmental beneficence, as where new legislation comes into force that favours a product.

2. Incremental opportunism, adding services or products in direct response to customer requests, in an incremental and reactive way.

3. Developing a related package of services and products, in a more deliberate attempt to link diversification, possibly by setting up related companies.

4. Actively searching and targeting additional niches that have different demand characteristics, thereby spreading risk.

5. Building a flexible resource base so that the firm becomes capable of supplying a wide variety of customers and products.

6. Applying expertise in related product areas, through a developed design function and radical product innovation.

7. Testing markets and boundaries, by experimenting with new business, including international ventures, and mixing elements of strategy (Hendry, Arthur and Jones, 1995).

These strategic options have all been discussed within Part 2, but the act of gathering them together (as well as providing a useful checklist) brings out the extent to which they can be distributed along the line that Mintzberg and Waters describe, which has "deliberate" and "emergent" as its two endpoints:

> Strategy formation walks on two feet, one deliberate, the other emergent . . . Managing requires a light deft touch — to direct in order to realise intentions while at the same time responding to an unfolding pattern of action. The relative emphasis may shift from time to time but not the requirement to attend to both sides of this phenomenon . . . (Mintzberg and Waters, 1985).

The most deliberate strategies, for Mintzberg and Waters, originate in formal plans that are clearly defined by management, and that enjoy structured implementation in a stable environment. Different points on the line between which runs from deliberate to emergent include "umbrella", in which the organisation operates within certain constraints, that oblige management to define the strategic framework in which individuals do their best to cope with a complex environment. While the most emergent strategy is described as "imposed", in which the environment dictates the actions and pre-empts organisational independence of decision and action.

This model highlights the difficulty which many businesses encounter in distinguishing between strategy and objectives. Ansoff, who defines strategy as "an elusive and somewhat abstract concept", appears to provide a clear distinction:

> Since both strategy and objectives are used to filter projects, they appear similar. And yet they are distinct. *Objectives represent the ends* which the firm is seeking to attain, *while the strategy is the means to these ends*. The objectives are higher-level decision rules. A strategy which is valid under one set of objectives may lose its validity when the objectives of the organisation are changed (Ansoff, 1965).

But the following paragraph, quite rightly, demolishes his own carefully-constructed house of cards:

> Strategy and objectives are interchangeable; both at different points in time and at different levels of organisation. Thus, some attributes of performance (such as, for example, market share) can be an objective of the firm at one time and its strategy at another. Further, as objectives and strategy are elaborated throughout an organisation, a typical hierarchical relationship results: *elements of strategy at a higher managerial level become objectives at a lower one (ibid.)*.

Despite all the theorising, however, our conclusion brings us back to the necessity for the successful manager to be able to harness the resources of the organisation to bring about appropriate change. The particular skills that this demands are those that Schein attributes to the process consultant:

> The process consultant seeks to give the client insight into what is going on around him, within him, and between him and other people. Based on such insight, the consultant then helps the client to figure out what he should do about

the situation. But the core of this model is that the client must be helped to remain "proactive", in the sense of retaining both the diagnostic and remedial initiative (Schein, 1988).

As Schein observes, the objective, for both consultant or manager, is to prevent the (internal or external) client from becoming dependent:

> Implicit in this model is the assumption that all organisational problems are fundamentally problems involving human interactions and processes. No matter what technical, financial, or other matters may be involved, there will always be humans involved in the design and implementation of such processes. A thorough understanding of human processes and the ability to improve such processes are therefore fundamental to any organisational improvement *(ibid.)*.

In an organisation which is developing and implementing an innovative project, there will be networks of people engaged in achieving common goals, and thus numerous processes occurring between them. The greater the skills and resources that are deployed to understand how to diagnose and improve such processes, therefore, the more likely it becomes that solutions will be found to technical problems, and that these solutions will be accepted and used across the organisation.

Equally, the more radical or complex the innovation project, the more likely it becomes that political behaviour needs to be taken into account at all stages:

> Within decision-making processes power strategies are employed by the various interested parties through their demands . . .

> Not all demands can be met. In the absence of a clearly set system of priorities between those demands, conflict is likely to ensue. The processing of demands and the generation of support are the principal components of the general political structure through which power may be wielded. The success any claimant has in furthering his interests will be a consequence of his ability to generate support for

his demand. The final decisional outcome will evolve out of the processes or power mobilisation attempted by each party in support of its demand . . .

In an innovative decision process involving executives and change agents, the issues that are likely to arise will have to do with the relative contribution that either side can claim for its knowledge or skill contributed as resources, and the right thereby to the greater or lesser share of command over total resources. The generalised demand by the innovators is likely to be for increased recognition of the importance of technical information as a business resource, and therefore for increased standing as the controllers of such information. This may be seen by executives as a demand for quasi-elite status. It often is (Pettigrew, 1973).

For all Pettigrew's insight, however, probably the best book on organisational behaviour was written 500 years ago (although not published for a further 40 years):

It should be borne in mind that there is nothing more difficult to arrange, more doubtful of success, and more dangerous to carry through than initiating changes in a state's constitution. The innovator makes enemies of all those who prospered under the old order, and only lukewarm support is forthcoming from those who would prosper under the new. Their support is lukewarm partly from fear of their adversaries, who have the existing laws on their side, and partly because men are generally incredulous, never really trusting new things unless they have tested them by experience (Machiavelli, 1532).

The objective of any architect of change is thus to reduce the risk of "new things" to a minimum through following what may appear, at a first reading of Part 2 of this book, to be a lengthy and complex process. Which it is, since although some of the stages can be reduced in some firms (due, for example, to a broad and detailed knowledge of the current state of the technology), none can be ignored. But, just as the risk of not innovating invariably outweighs the risk of innovating, so the risk of telescoping the innovation process significantly increases what Machiavelli describes as the dangers implicit in change.

PART THREE

# WHY SMALL FIRMS ARE SPECIAL

*Chapter 11*

# INTERNAL STRENGTHS AND WEAKNESSES OF SMALL FIRMS

Contents

*This chapter and the following one represent a break with the shape of the book up to this point. In them we define what it is that distinguishes small-medium firms, in so far as the process of innovation is concerned, from multinational and multi-product giants. There is a fundamental distinction to be made between the two; and a coherent strategy can only be built upon an understanding of what it is that is different between one business and another.*

*This chapter will be devoted to the internal characteristics of small-medium companies; and the following one to the relationships between the business and its environment. But, although all businesses are dependent on the same economies, and the same marketing theories, for example, apply to each of them, the differences between small and large firms cannot be understated. A small company based in Perpignan in the south of France, for example, has more in common with the local town or regional administration than it does with a multinational like Aerospatiale, even if it is one of their sub-contractors.*

*This book has chosen to address in particular the issues which affect the managers and other decision-makers in small firms. This is why it is necessary at this point to highlight the structural characteristics of this type of business and to bring out some of the consequences of this on the processes which involve and advance innovation.*

*The traditional definitions often look upon large organisations as being efficient, and small ones as something that has not quite worked out. But these views no longer have any validity, due in part to the changes in the rules on competition in local, national and international markets. Because of this, small companies have won a significant share in many markets. In this context, any entrepreneur who is able to identify the internal strengths and weaknesses of her business can have a significant impact. Besides, the readjustment which is currently taking place in a number of large companies as a result of the competitive demands of more open markets is creating new opportunities for smaller companies to exploit. These include:*

- *The ability to respond to changes in customer needs*

- *A new division of labour between small and large companies*

- *Internal and external flexibility*

- *Flexible specialisation.*

*The entrepreneur is therefore faced with a number of possible scenarios. Appropriate strategies for each of these are described below.*

## 11.1 The Advantages and Disadvantages of Small Firms

One never defines anything as clearly as when contrasting it. Popular business opinion has for a long time considered the smaller firm, when compared with the large company, to be some kind of inferior version of the real thing. It is certainly true that the small firm has a distinctive range of organisational forms. The processes are much simpler, more fluid and more personal; while the procedures are less standardised, and internal communication is more easily controlled; additionally, the individual influence of the entrepreneur is critical. For a long time, this type of organisation was considered to be a weakness. It was synonymous with a greater exposure to uncertainty and risk for the owner, to financial vulnerability and to uncertain employment prospects for the workforce. What all of this implied was that an efficient small business should, over time, move towards becoming a large business.

This image changed substantially in the course of the 1970s and 1980s, due both to the exceptional performance over this period of smaller companies in almost all the major industrial economies, and to the crisis which hit the large companies with much-publicised waves of redundancies. These were in many cases followed either by real bankruptcies or by those concealed behind public rescue plans (dockyards, banks, the steel industry), and major financial mergers or substantial take-overs. It was through all this that the potential for adaptation and flexibility of the small firm was rediscovered (for, as Schumacher argued, small sometimes can be beautiful); and, through this, there has been a revival of production in batches or in small runs; and of the capacity to produce a steady stream of small innovations, as well as a number of revolutionary innovations. The value and originality of the management of a small business has even been recognised, given the ways in which it can be seen as a potentially creative force and as a stimulus to innovation. Finally, the issue of job security is now considered in a more dynamic light, as the opportunity to adjust the balance between labour supply and demand. Everyone accepts that this flexibility has its darker side as well, both for the employees and the entrepreneur. But this in itself is one of the characteristics of smaller organisations.

It is essential however to emphasise the opportunities and barriers to be found in the areas of human resources, innovation, investment and marketing.

### *Human Resources — Recruitment, Management, Training*
Human relationships take on a very distinctive character in a small company. There is little bureaucracy, little real hierarchy, and the amount of direct contact which occurs can increase a sense of responsibility and initiative. But, at the same time, the uncontested power of the owner-manager (that is to say the absence of counter-balance), and the gradual concentration of debate around less important production issues, conceal more essential differences of opinion and the need for radical self-examination, because professional behaviour and individual behaviour are so easily mingled and confused. Like any partners who have been

together for a long time, directors and employees often grow older together and communicate less and less.

This means that the smaller company (where it organises itself to make the most of this) is better adapted than the large to make the most of the advantages to be gained from flexibility in the labour market, both internally and externally. Internally, the division of tasks can be less precise, and individuals can more easily move around, with training in new skills being possible on the job. Externally, however, the opportunity to move is very much dependent on the state of the local job market. It is harder for a small business to find "good" people, and still harder to develop them inside the organisation. In this case the size of the business makes career planning difficult, as well as limiting the ability of the business to recruit a broad range of skills. Further disadvantages are that it is perceived to be difficult to develop close links with a university, for instance; and, most important of all, a small firm is less likely to be able to assemble an attractive financial (and benefits) package. In the past, some of the advantages which large organisations enjoyed were counteracted by lower employment costs in smaller companies, but this is no longer the case: whatever the size of a business, it now needs to hire professional specialists (for IT systems, setting up specific projects, etc.).

Figure 11.1 summarises the strengths and weaknesses of small firms where the management of human resources is concerned. From this it is clear that each strength is also a potential weakness, and vice versa.

## Figure 11.1: Human Resources in Small Firms

|  | **Areas of Strength** | **Areas of Weakness** |
|---|---|---|
| Recruitment | • Easy movement between businesses<br>• Rapid process based upon a "feel" for people | • Dependence on local labour markets<br>• Highly subjective |
| Management | • Functional and flexible | • Little career planning |
| Training | • Learning by watching and imitating ("learning from Fred") | • Little investment in training (with the exception of specific training for immediate use) |

Company behaviour with regard to training is illustrated by a study carried out for this book in Italy (the Lombardy region) and in France (Île-de-France) using a range of innovative companies. This study enabled a classification to be made of the obstacles to training in small firms. The first obstacle is that of cost (which is mentioned by 50.4 per cent of businesses in Italy and 49.6 per cent in France). The second obstacle is the time wasted on travel from the business premises to the training venue (37.8 per cent in Italy, 39 per cent in France). The third reason given was the inability of the training on offer to cover changing individual needs (29 per cent of businesses in Lombardy, and 21.4 per cent in Île-de-France). The least significant obstacle was felt to be the poor quality of the training provided (28.7 per cent in Italy and only 14.8 per cent of the French businesses). Other research in West London shows that British small firms have views on the obstacles to training which are very close to those of their French equivalents, although a significant number of West London companies (over 20 per cent) listed uncertainty over what type of training would be appropriate as a significant obstacle. It is perhaps worth mentioning in passing that the importance attached to these issues varied considerably from sector to sector.

Beyond the issue of training is to be found that of rapid changes in behaviour. When a business is much more dependent on quality than on price, a more participative human resource policy is called for. The rules of partnership and the allocation of responsibilities change, and the reward systems need to do so as well. In this case, the flexibility of the labour market makes vigilance essential. An excellent employee who is frequently praised but only paid at the same level as less effective colleagues may not stay in the business for too long. This is why there are an increasing number of examples of employees who have received a bonus of some kind as a reward for suggesting a significant innovation. Nor should someone who comes up with a range of minor innovations fail to be rewarded in the same way.

### Innovation — Incremental and Systematic
In this area as well, small firms are different. They possess a great capacity for unplanned innovation in their everyday activity, around issues like developing new processes. But what they are

less able to do is to direct the necessary resources at following this idea through to its logical conclusion and impact. This is because small firms are invariably less effective when it comes to planning and implementing innovation, or to establishing a formal technological research group or R&D department, or to allocating specific individuals to work exclusively on developing innovations.

This does not however mean that small firms cannot develop:

- Project teams able to realise the total development, production and launch of a product

- Teams with a high level of technical expertise and specialising in technology watch (see Chapter 13)

- A dynamic range of technical skills (even if these are not part of the formal structure) spread across the organisation in a number of employees.

Nor should it be forgotten that process improvement is not the same thing as process innovation.

**Figure 11.2: Process Improvement and Process Innovation**

|  | Incremental Improvement | Radical Innovation |
|---|---|---|
| Starting point | Existing process | Clean slate |
| Frequency of change | One-time continuous | One-time |
| Time required | Short | Long |
| Participation | Bottom-up | Top-down |
| Typical scope | Narrow, within functions | Broad, cross-functional |
| Risk | Moderate | High |
| Primary enabler | Statistical control | Information technology |
| Type of change | Cultural | Cultural/structural |

The study referred to above revealed that in Lombardy the source of an innovation more often than not is external rather than internal. Data from 460 businesses shows that internal processes (through specific R&D) only resulted in an innovation in 26.8 per cent of cases; whereas external factors (in particular customer needs) accounted for the other 73.2 per cent. There are significant differences however between the behaviour of manufacturing as opposed to service organisations. Internal innovations represent 36.1 per cent for the former, as opposed to only 11 per cent for the latter.

*The sources of innovation thus tend to be more likely to be unpredictable and external in the case of smaller than for large organisations.* Nothing guarantees, for example, being able to file a precise number of patents. But the innovations in small firms reveal a much greater variety, since they are the result of a much more informal, and potentially much more creative, organisation and process.

It is thus possible to put forward the argument that small firms are, at least where their organisation is concerned, well-adapted to generate innovations within clearly defined technological boundaries. Their aim is frequently to know how to refine, improve and vary technologies and products, rather than to reach radical new goals. In this situation, the handicaps which result from reduced and unsystematic R&D activity certainly diminish the capacity of small firms to extend boundaries and to identify new technological opportunities. There is no doubt however that large industrial groups have grasped the fact that, despite a lack of resources, small firms are an inexhaustible source of innovations. In many cases, their strategy is to keep a close watch on innovative small firms, rather than on emerging technologies. They even go so far in some cases as to specialise more in gaining maximum commercial benefit from innovations, than in their creation (see the Boston Consulting Group model in Chapter 5). Bearing this in mind, it is possible therefore for a medium-sized business to "grow from seed" something that a large business will have to invest very heavily to acquire.

In many small firms, however, the process of innovation tends to leave too much to chance, and so evaluation is rarely given the significance that it deserves. The Lombardy study shows that 50.8 per cent of small firms have no mechanism in place to evaluate the real costs of their investments; and this percentage rises to 67.4 per cent for businesses in the service sector, against 38.6 per cent for those involved in manufacturing.

In other words, the small firm needs to learn how to make innovation both a *state of mind* and a *habit* (and not just a new word for improvement). It is not possible, in the long term, to leave this crucial motor for change in the hands of spontaneity or of chance. This is a fundamental area of weakness in the majority of small firms, but it can be managed through the use of a series of measures which make access to technology and to science less alarming, by introducing a change culture. *Every business needs to become a learning organisation*. Over time, a change culture can move a business from a dependence on the brilliant one-off discovery towards one in which innovation (in the broadest sense of the term) enables the business to achieve a greater penetration of its target market.

**Figure 11.3: Innovation in Small Firms**

---

**Areas of strength**

- Flexibility

- Routine procedures can be reduced to a minimum

- A more rapid movement of ideas around the firm.

**Areas of weakness**

- Unsystematic approach to innovation

- Innovation often happens by luck or accident

- Short life span of most innovation groups

- Mistakes can have serious consequences (burden of responsibility).

---

### *Financial Investment*

The issue of financial strategy is obviously a critical one. Successful entrepreneurs are often excellent technically, as well as being specialists who are very interested in the production process and in new products. An entrepreneur is often also an expert in combining technologies and products, but only rarely possesses a strong background in finance. There are obviously exceptions to this caricature, due either to the particular skills of one individual or to the range of skills within a management team, one of whom may be a specialist in production or marketing, and another in finance (this is not a rare combination, even in family-run businesses).

This is however the exception, and as a rule individuals who are lively entrepreneurs tend to have a weakness on the financial side. For example, a recent study in Modena (an Italian area that is particularly rich in entrepreneurial small firms) showed that only 22 per cent of businesses have a formal financial function, whereas 97 per cent had a formal administrative function, 78 per cent had stock control, 74 per cent had a sales function, 66 per cent had general management, 62 per cent had personnel, and 55 per cent had forecasting.

This situation is not improved by the tendency of financial institutions (and notably banks, although this varies from region to region across Europe) not to pay much real attention to small firms. The small period of time allocated to each customer, and the difficulties associated with assessing a true (as opposed to a worst case) level of risk means that small companies invariably receive a lower level of attention (notably at the strategic level) than do their larger counterparts. Faced with a lack of direct and collateral guarantees, banks tend to be over-demanding in terms of security, and this has an immediate and restrictive impact on any innovative project for which the small firm is seeking funds.

Financial handicaps, however, are not only due to the banks. Government also has a tendency to penalise the smallest businesses. Access to state funding for innovation favours those larger businesses able to spare the resources required to jump through a number of bureaucratic hoops, without which support cannot be granted. And the more a bureaucratic system is centralised, the more likely is it to favour structures as large as itself.

The owners of small firms do, however, have to accept some of the blame themselves for these difficulties in the area of finance. One comes across too many cases where the entrepreneur uses his own lack of financial expertise as an excuse for not taking a financial risk. Not resorting to borrowing becomes a point of honour, linked perhaps to a preference for possessing 100 per cent of a £300,000 business rather than 50 per cent of a £3,000,000 one. But there are no shortage of borrowing opportunities available and arrangements are possible for any business wishing to strike a balance between its own resources and those of a wide choice of financial markets and services. In this field, advice from professional bodies, chambers of commerce, accountants, suppliers and competitors is often invaluable. There is equally a range of "new" financial services which should be investigated, such as credit unions (as used in Italy in particular).

**Figure 11.4: The Finance Function in Small Firms**

**Areas of strength**

- Speed of response of the entrepreneur

- Growing number of financial tools available

- Increasing competition between banks.

**Areas of weakness**

- Lack of competence when it comes to financial analysis

- Absence of financial strategy

- Financial markets hard to access without having capital oneself

- Difficult to recruit good financial managers

- Vulnerable to late payment by larger firms and public bodies

- Difficulty accessing grants and subsidies.

*Management of Stock and of Channels to the Market*
The problems associated with the management of stock, transport and storage space are often also substantially different for smaller businesses. The smallest are often faced with not really having a way of satisfactorily developing these functions. But information technology can often be used nowadays to reduce the problem to a much more manageable size.

There is often little distinction made between the areas of stock and transport and that of marketing. The processing of orders and sub-contracting can be used to integrate the logistical function as a whole. The buying function (sometimes the purchasing department is dependent on the technical and manufacturing arm) is often closely linked to personal behaviours and to individual reactions.

Figure 11.5 below identifies the key channels to the market.

## Figure 11.5: Key Channels to the Market for a Small Business

**Direct channels**

- Direct contact with business customers

- Specialist agencies

- Public tenders

- Representatives in the domestic market

- Agents or dealers in overseas markets

- Export offices overseas.

**Indirect channels**

- Major retailers

- Wholesalers and importers

- Working for businesses who sell on into their own market

- Import-export groups.

Naturally, the position of a particular business will depend on the economic, territorial and organisational activity in which it is involved. In every case it is a question of choosing between different channels, and, where required, of using several channels at the same time. There is obviously a link between diversification into other channels and the size of the business.

The most straightforward and traditional strategy is that of a principal customer who ensures a regular flow of orders. A large business, where it is a sole purchaser, ensures the orders and defines what is very much a dependent relationship. The subcontracting business is bound to the provider of the order in the area of quantity, technology, quality, price, and therefore of profits. But when for instance the large business is going through a difficult period (due to internal or external factors) the subcontractor is inevitably, and without having much say in the matter, used as a shock absorber, and thus experiences the effects in an intensified way. For this reason, it is important that subcontractors take steps to increase the number of businesses for whom they produce. Through doing this they reduce risk, and can also pass on to the others the benefits of progress made in working with one customer.

Another strategy — and a highly effective one — is to develop a *market niche*. This involves developing a product (and the offer which surrounds it) which, by virtue of its own features, is able to defend itself against any competitor in the sector, whatever their size. Personalising the product and the addition of appropriate options is all part of this strategy. Some businesses specialise in producing something complex exclusively for one customer, but in this case the level of skill involved and the impact in the marketplace mean that they are a long way from being a dependent subcontractor. Nevertheless, there remains a real dependence on the success of the customer's business.

A number of businesses in Reggio Emilia illustrate this type of strategy. In the sector containing pumps and equipment for watering and irrigation, five small firms in the area represent 67 per cent of the total Italian market. In addition, over 50 per cent of their output goes for export. In the diesel engine sector, three local producers (of whom two have between 300 and 500 employees) have 85 per cent of the national market between them.

Instead of concentrating on one or two niche products (which, like any other product, have a life cycle which cannot be indefinitely extended) the business can make the decision to extend its product range, sometimes considerably (components and accessories), or to produce a catalogue. In this last case, the channels used for distribution are diversified, and a greater balance achieved between other producers, dealers, multi-product representatives, supermarkets or subsidiaries.

In this way, the small business can exploit its potential for flexible specialisation.

## Figure 11.6: Flexible Specialisation in the Market

---

**Areas of strength**

- Opportunities provided by "just-in-time"

- Possibility of varying market strategies

- Effectiveness of highly focused niche strategies.

**Areas of weakness**

- Poor knowledge of the market

- Management of sub-contracting relationships

- Some niches are dead ends.

---

### 11.2 New Scenarios

There has been no shortage of studies of the changes which many large companies have gone through, including the way in which many of them have become groups, thus combining the benefits of financial muscle, of being a multinational and of decentralised production on a human scale (which is one of the key advantages which smaller companies enjoy). Olivetti, Fiat, Saint-Gobain and Rhône-Poulenc are no longer monolithic structures. At the same time, they have rediscovered the advantages to be gained from a manufacturing business having local roots. They have also come to appreciate the benefits of developing a new spirit and networks of co-operation with their competitors, and also with the key players in the local economy, notably independent smaller businesses. These recent changes have had a real impact on the opportunities which are available for small firms, who now have to come to terms with new forms of demands, risks and competitive environment.

It is with three particular opportunities that this chapter will conclude.

Firstly, the ***creation of new businesses*** which specialise in serving the small firms sector. These are due to a number of specific factors:

- The increasing integration of existing groups, and their new strategies in forming local partnerships outside of the group itself, through subsidiaries which specialise in services to business; this is particularly visible in the areas of spin-offs and development.

- The development by the banking system and by new forms of financial intermediary of services for cash management and for support measures accompanying high-risk investments.

- The rethinking, not a moment too soon, of public policy towards supplying real services rather than towards financial incentives (this is what lies behind the regional policies of the European Union, as opposed to financial support for the member states).

The second area of opportunity is the ***creation of a range of totally new services*** by the subsidiaries of large manufacturing companies, to satisfy the new range of demands which small firms are making.

The third area of opportunity involves the ***development and strengthening of stable local supplier networks,*** in areas where in the past the suppliers have been relatively autonomous.

The distinction which needs to be made between the second and third of these lies principally in the extent to which autonomy is retained. Where this remains high (as in the second scenario), a medium-sized business retains the competitive advantages which have enabled it to develop in the past and to establish a mutually beneficial relationship with the larger players. In the third scenario, on the other hand, it continues to develop essentially as a decentralised production unit enjoying the social benefits of decentralisation, but in a more dependent relationship with larger companies, with regard to the flow of information and to its relations with the market.

There are a number of possible responses to these three scenarios:

In the first case (the creation of new businesses), some businesses will make good their weaknesses in key strategic areas through strengthening their internal functions in management and in innovation planning; and also through achieving integration between product/process innovation and the organisation (i.e. seeking to achieve effectiveness as opposed to efficiency in innovation).

In the second case (the creation of a range of totally new services), companies will be able to participate in the creation of a network of real services, either independently or within the context of the activity of regional or local economic development agencies.

In the third case (the development and strengthening of stable local supplier networks), companies will collaborate with the large organisation through negotiating terms which will allow them the greatest possible autonomy.

This whole discussion, however, needs to be seen against a background of what (for want of a better description) can be called the "Ideal European Small Firm".

## Figure 11.7: The "Ideal European Small Firm"

The "Ideal European Small Firm" will be different from most of its local counterparts through its activities and aspirations. These include:

- Management of the entire business around a common project

- A significant proportion of turnover in exports (often using new products)

- Exploiting a niche market

- Highly professional management style

- Effective use of advanced technology mixed with well-tested technologies

- Not being restricted to local sourcing

- Aptitude at networking (with a view to obtaining new product ideas and technical information)
- Pursuit of competitive alliances
- Fostering long-term relationships
- Willingness to share equity with external individuals or organisations
- Willingness to introduce new products
- Willingness of owners to devolve decisions to non-owning managers.

*Chapter 12*

# THE ENVIRONMENT

*It is when the environment is discussed that the difference between large and medium or small businesses is seen at its most marked. The large business or an industrial group is frequently able to have as much influence on its environment as on its own structures. The small firm, on the other hand, is in a situation in which it is highly dependent upon this same environment. In theory at least, it has no power over the market. Rather it is the market forces which will determine its behaviour. For a small firm, there is no monopoly, fewer economies of scale, and, in the majority of sectors, barriers to entry which are more easily overcome by its larger competitors.*

*But fortunately this does not mean that the small firm is entirely dominated by its environment. It can benefit from a greater freedom of movement than is often imagined to be the case. This room to manoeuvre is the subject of this chapter.*

*There are a number of very different situations depending upon whether the business has secured a niche or whether a large number of competitors exist; upon whether the competitors are of similar weight or whether a particular group dominates; upon whether the competitors have offensive strategies or not; or upon whether the sector is technologically stable or whether it is a fast-moving and very "high-tech" area. It is all a question of making a comparison between the different players: a comparison of assets, of order books, of behaviour and of strategies.*

*There has certainly been one significant recent development. It is increasingly in the interests of a dynamic business to establish links with its environment — but obviously not any links at any price. The traditional opposition has been between, on the one hand, the modest, self-effacing and stable business, and, on the other, the profitable and far-seeing business, whose managing director is always popping up in the local media, and who receives the "Manager of the Year" award before her company is acquired by a major group in the course of the following winter. But this is no longer an accurate picture of the situation in which small firms find themselves. Beyond these two caricatures is a different form of business existence, built upon the systematic creation of flexible relations with different partners: the large company which provides the order, and a range of small firms, or the public and private sector specialist agencies. Establishing these networks does not eliminate competition. but it certainly changes its nature, and generates new opportunities to explore and exploit.*

## 12.1 From Dependence on the Environment ...

The normal situation for a small firm is to be dependent on its environment, but there is no shortage of sources of information to enable a business to keep abreast of external pressures (see Figure 12.1).

The environment is invariably the strongest constraint, today more than ever before, on the small firm dominated by the market. Prices are fixed by the competition, often by businesses with considerable muscle and against whom little can be done. Consumers are becoming more and more demanding. Intermediaries are also behaving in an increasingly competitive way, refusing to be tied to a supplier. Unless one is lucky enough to produce something that is unique and highly sought-after, market conditions fix the rules of business activity.

## Figure 12.1: Gathering Information

There is no shortage of information that is readily available (either at no charge or at low cost). The following list summarises the key sources, and identifies the coverage of each source.

| | Competition | Regulations | Technology | Market |
|---|---|---|---|---|
| Business media | X | X | X | X |
| Business organisation | | X | | X |
| Company accounts | X | | | |
| Competitor products | X | | X | |
| Conferences/ Seminars | X | X | X | X |
| Consultants | X | X | X | X |
| Customer contacts | X | X | X | X |
| Euro Info Centres | X | | | |
| Foreign markets | X | | X | X |
| Government trade and industry dept. | | | | X |
| Kompass | X | | | X |
| Market studies | X | | | X |
| Patent searches | X | | X | |
| Product literature | X | | X | |
| Regulation enforcers | | X | | |
| Staff contacts | X | X | X | X |
| Technical literature | | | X | |
| Trade asso- ciations/press | X | X | X | X |
| Trade shows | X | | X | X |

The second area of dependence is on large companies, or simply on those businesses which dominate the market (and which are usually large companies). Size alone does not guarantee market power, but it does tend to be these companies which dictate the rules, and which determine advertising levels, quality standards, the moment to launch innovative products (which may have been kept in drawers for years), product design, etc. And it is important to follow immediately, without giving the impression that one is a follower. But the only realistic way to make up the gap with the competitor who has taken the initiative is to beat them on price, and so almost inevitably to condemn oneself to operate on tighter margins.

A third area is dependence on the law. Any law voted by parliament is universally applied by a mass of administrative organs. For any business, the proliferation of rules and regulations represents a permanent administrative burden. Nobody has a good word for bureaucratic red tape, but everyone is happy that there are laws to resolve conflicts, and roads to transport products, as well as the regulation of telecommunications or business tariffs for electricity. All of these are basic conditions for industrial development and individual entrepreneurship. That they are essential becomes apparent as soon as they disappear, as occurred in parts of Eastern Europe. In this respect, if the state did not exist, business would have to invent it.

The list of constraints with which any business must contend is a long one. It is certainly a source of stress but it equally demands an effort to behave differently, to rethink relations with partners, or to search out innovations. This is the logic of the market. It leads to a particularly high level of transaction costs, in the sense of being forced to negotiate each external operation where one does not benefit from integrated industrial structures.

In this area, the small business is considerably disadvantaged compared to the large. The latter has managed to "internalise" a large number of activities. The small business, on the other hand, continues to be totally dependant on the exterior for information on constantly changing raw materials prices, for purchasing inputs, for training its workforce in new technologies, and to make sure that its technology develops or that it has access to credit.

## 12.2 ... To the Development of Local Networks

As in other areas, the disadvantage which is described here is not an absolute one. If it were to be seen solely in the light of the previous section, every small firm would be entirely dominated by its environment and would either soon cease to exist or be trapped in a relationship of total dependence. But for most of the time, this is not the case. Why? Essentially through the development of a network out of a certain number of "proximity" relationships where, in the last analysis, each partner brings something indispensable to the table, so that the resulting synergy is greater than would be achieved individually, even by the most powerful partners.

Every area of business activity leads to the creation of a mass of stable relationships between partners who have very different goals and interests, but who share, on one point at least, a common interest. Proximity can be identified on a number of levels:

- **Geographical proximity,** which consists of a mass of daily exchanges which develop with immediate neighbours, the local council, customers, suppliers. This proximity will naturally be more significant and enriching where the locality is dynamic, diverse and complex, and looks to provide general conditions in which growth can occur (a good road system, high quality business support services).

- **Social proximity,** which is the result of a shared knowledge and experience. This occurs, for example, where business managers attend the same trade events (devoted to agricultural equipment, to vehicles, to office technology), read the same specialist publications (concentrating on polymers, machine tools, soldering technology), and are keen to share their knowledge and experience whenever they meet.

- **Technological proximity** consists of the differences and synergies which exist between businesses working in the same sector, with the same equipment, or with the same type of organisation. These proximities are strategic.

- **Productive and commercial proximity** often replaces the authoritarian power structures of the subsidiaries of major groups. It shapes the relations which are required in order to rationalise the different stages of the production process (supply chain, eventually to include subcontracting).

- Lastly, **regional and national proximity.** The opening up of frontiers has not necessarily led to the disappearance of political units. In fact, it has had quite the opposite effect. The impact of national industrial policies, but also of all those which exist at a regional and at an international level, naturally create living and effective proximities. Leaving aside the lure of the exotic, it is still easier for a British business to do business with another British business than with a business in Japan (although ease alone is not a reason for doing business).

**Figure 12.2: Partnership: The Business Needs to Be in the Centre**

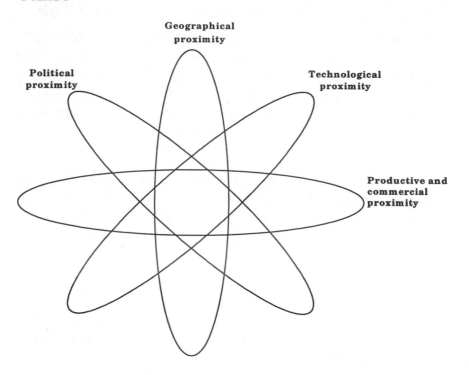

These proximities define the limits of two spheres of activity outside the business: a first sphere dominated by relations with the *market and competition*, and which finds its principal expression in the choice of customers; and a second sphere dominated by *partnerships* based on these different proximities. The ability of a business to establish contact with its close partners has become a key factor in its effectiveness, and even a condition without which it may find itself left behind. But the increasing complexity of the business environment can make this approach a difficult one: opportunities and difficulties often go hand in hand.

The following opportunities need to be considered:

- The extent to which the partners (whether public or private sector) know and understand each other's goals and development plans

- The extent to which a business has increased its own knowledge base, and thus its potential to generate innovative ideas

- The discovery of ways to add value to this knowledge base

- Extending activity beyond its own physical limits and being able to benefit from some of those advantages which are normally found in larger businesses (such as economies of scale), while retaining the traditional strengths of flexibility and organisation

- Lastly, collaboration enables a sharing of the expense and of the risk, especially for the most risky innovations.

The following difficulties need to be considered:

- The need to go beyond a relationship based merely on confrontation and competition in the marketplace

- Establishing new rules around issues of business confidentiality and how it will be ensured (confidentiality does not vanish, but its scale and nature needs to change)

- Those involved in creating a new form for the circulation of information. In a world in which information has become a commodity, the issue of its diffusion, circulation, and therefore of its ownership, is a strategic tool and a key area of potential conflict. It is therefore important to be able to assess the risks and the benefits of this form of interaction, and to act in consequence. One is faced less with a dilution or a loss of autonomy

or of personality than with a new and changing relationship
between independent businesses; and where this can be skilfully managed it is a game which both should be able to win.

## 12.3 Networks

It is now time to look at the impact of the infrastructure of the
local area (be it a strong area or a weak one), at the value of the
direct or indirect relations between the locality and the different
players operating there, and finally at the relations with the large
business that is viewed as an economic reference point. It is important to note that it is the role of any business to participate in
the creation of production and innovation proximity systems.

The creation of these proximity systems obviously depends on
what kind of structures already exist locally. Clearly, a major influence on local dynamism is the presence or absence of specific
resources in the area: the proximity of a major population centre;
the presence of organisations specialising in scientific and technological research; the presence of businesses offering highly specialised services, or of technopoles. The existence or otherwise of
these elements enables more precise choices to be made in establishing those contacts and relationships which have most to offer
to a firm's own strategies. A ***strong territory*** is one which contains a high concentration of these organisations and agencies.

---

### A Strong Territory:
### Baden-Württemberg (BW)

*Large companies play a decisive role in this region of Germany, in
the electronic and computer sectors as well as in the car industry.
But the region (and the city of Stuttgart in particular) contains a
large number of small firms which are subcontractors. Because of
this, small firms employ over half of the manufacturing workforce,
with businesses employing between 50 and 500 employees being a
key factor. This is therefore a manufacturing infrastructure with
low levels of concentration.*

*The strength of the small firm sector is a major factor in the
economic strength of the region as a whole. Their key characteristic*

*is the combination of flexibility and specialised production . . . and the creation of wealth and employment.*

| Manufacturing Businesses in Baden-Württemberg (more than 20 employees) | | | | |
|---|---|---|---|---|
| Size | Number of Businesses | % | Total Employed | % |
| 500 and over | 455 | 4.90 | 710,282 | 49.4 |
| 50 - 499 | 4,222 | 45.4 | 600,516 | 41.7 |
| 20 - 49 | 4,642 | 49.7 | 127,470 | 8.90 |

*The table below shows that these businesses do not merely provide support for the large firms in the area, but represent a technological core which has direct access to national and international markets.*

| Percentage of Turnover from Exports (by Industrial Sector and Company Size) | | |
|---|---|---|
| Sector | Number of employees | |
|  | 100 - 499 | 500 and over |
| Mechanical engineering | 43% | 50% |
| Electronics | 27% | 34% |
| Chemicals | 31% | 39% |
| Precision engineering, optics | 43% | 4% |
| Textiles | 23% | 30% |
| Plastics | 24% | 27% |

*The development of this highly-qualified and independent industrial base has been assisted by an **effective system of innovation grants**, by the **availability of expert advice** and by **co-operative research** involving a number of local organisations (university, chambers of commerce, trade associations, foundations). These activities represent 25 per cent of the total volume of technology transfers. The remaining 75 per cent take place between manufacturing companies. This demonstrates a high level of partnership based on co-operation, rather than pure opportunism, within the local economy. This attitude has a crucial influence on the flexible and specialised nature of local companies, and ensures a significant level of dissemination of technological knowledge as* *opposed to individual businesses keeping their secrets in locked drawers.*

*There is one other significant factor which contributes to the success of Baden-Württemberg: a large number of collective salary agreements, particularly with regard to qualifications and training.*

---

A **weak territory**, on the other hand, will result in a different set of problems. Difficulty in establishing relations with the exterior, due to a shortage of structures suited to a local economic development policy, results in a slow growth of activity, and to individual businesses being isolated, as well as to a lack of confidence in the possible benefits of working together.

The terms "private" and "public" are general labels that are given to the different players operating in an area. Within the private sector customers, suppliers and competitors are very much key players, who have traditionally had a very specific relationship with the business. But the private sector also contains agencies charged with the spread of R&D, advanced services and maybe even a policy on training the workforce. In the public sphere, as well as the regional and local administrations, there are a number of players who are becoming increasingly influential, such as the universities, the specialist colleges and research institutions; nor should social and trades union organisations be ignored, since they represent a key element in the social consensus that is necessary if industrial growth is to be developed and sustained. Equally, it would be an error to neglect the potential role of professional networks and associations in this process.

Collaboration between central and local government and other forms of association can lead (and there are examples of this in a number of regions in Europe) to a transformation in industrial terms: the creation of structures for training and for the diffusion of technological innovations, and social services in support of the workforce. A whole range of measures exist which need to be studied and used: grants, loans, development plans, but also economic policies appropriate to the growth needs of each area (labour market issues, skill shortages, regeneration).

It is obvious that this degree of local integration makes demands which may appear difficult to take into account from the very start. But all the structures that are already in place are not automatically partners. The environment is made up of players among whom the business must identify those which present the most suitable opportunities for its own development.

**Figure 12.3: The Direct Environment**

| The Business | | | | |
|---|---|---|---|---|
| • Customers<br>• Suppliers<br>• Competitors | • Social and trade union organisations | • Employer organsiations | • Private service structures (finance, production, training, R&D, legal advice) | • Public structures (local agencies, schools, universities, R&D agencies, regional agencies) |

To gain maximum impact from this approach to regional collaboration, it is vital to work to develop the idea of a network which takes the form of a kind of collaboration between different businesses. It may consist of a number of agreements and alliances, but also of relationships between different autonomous "business units" which are run according to the same rules, often based upon a particular business or guiding institution. The quasi-market, in the sense of a collaboration between businesses, or the quasi-business, meaning a number of relatively autonomous units, come under the general heading of network.

The following networks can be identified:

- Networks of external units, when the guiding business determines relations with the other businesses and with other structures

- Networks of internal units, where it is the internal organisation of the business which transforms itself with regard to hierarchy and relations, achieving a formal structure which resembles the external networks

- Interpersonal networks, where formal and informal relations influence the way in which the organisation of the business functions

- Virtual business, in which a business places an order with another business, which does not carry out the work itself but places the work (or a part of it) with yet another business. The provider of the initial order does not know, therefore, who is carrying out the work.

The network has a considerable capacity for developing and implementing strategy. From the point of view of competitiveness and competition, economies of scale can be achieved through the sharing of research arrangements and of product distribution channels. This divides and thus limits the relative costs, as well as offering the opportunity directly to control the technological development processes that are necessary to the business. The frontiers of the business are expanded, since the key aspect of the network is made up of the interactions between the different businesses, and no longer (or not only) of their internal structures.

---

### *The Example of Modena:*
### *How the Structures Are Linked*

*In this region of Italy there is a strong relationship between productive structures and the territory. This consists of an intense interaction between businesses, associations, and local government.*

*As far as the public sector is concerned, as well as grants to encourage new productive arrangements (the craft villages), there also exists a close collaboration involving, for example, the offer of services to business, while the chamber of commerce co-ordinates its involvement with the local administration and the regional government. Local government offers, as well as its social services, advanced services. The following have, for example, been set up for the crafts sector:*

- *An information bank on production in the clothing sector*

- *An observation centre on the provision of raw materials to ensure a greater fit between supply and demand from subcontractors.*

*With regard to training, professional associations and local government organise joint action in the following ways:*

- *An association for technological progress, created to train entrepreneurs, managers, and middle managers of businesses involved in working with metals and machinery*

- *A business centre involved in delivering training qualifications, concentrating on advanced services for businesses.*

*These initiaives demonstrate the extent to which a solid network of association between businesses, local government and the chamber of commerce is able to stimulate development and improve the introduction and use of innovations in private businesses.*

# Readings and Arguments 3

This essay covers Part 3 of this book. As with the first two essays, it is a critical exploration of recent and influential writing in the field, with a view to introducing the interested reader to further areas of reflection and action. Since the introduction to this book defined what is distinctive about small-medium firms, this essay takes a broader look at the environment in which they operate, beginning with the so-called "re-emergence of small-scale enterprise".

However, as with the first two essays, the third begins with a summary of some of the key arguments contained in Part 3.

---

### Key Arguments

- Small-medium businesses are not just large businesses in miniature, or the result of failed attempts to grow.

- The crucial strengths of these businesses are their ability to be both flexible and specialised, and to respond rapidly to changing customer needs.

- Successful small businesses do not necessarily want to grow in size or to measure success in terms of numbers employed.

- They need to play to their natural strengths (and to be aware of their natural weaknesses) in the areas of innovation, human resources, investment and marketing.

- Process improvement and process innovation are often confused, but are fundamentally different. The former is incremental, the latter radical.

---

- Small firms are an inexhaustible source of innovations, but are less effective when it comes to planning and implementing their exploitation.

- They need to learn how to make innovation both a state of mind and a habit.

- It is essential to be involved in changing the nature of competition through systematically creating and sustaining flexible "proximity" relations with different partners.

- A "strong territory" (dynamic local economy) displays high levels of partnership based upon co-operation, rather than on opportunism.

- Networks take a variety of forms and have a considerable capacity for developing and implementing strategy.

One of the theories that explains the re-emergence of small scale enterprise takes the view that the world's most advanced economies have now entered the fifth Kondratieff cycle, or long wave. In Britain, for example, the first long wave covered the period from around 1780-1840, and was powered by innovations in the production of iron, the mechanisation of cotton and the development of steam power. The second covers the second half of the nineteenth century, and is seen as having been based upon innovations in transport, especially the railways, the production of steel, and the extraction of coal. The third wave, lasting from around 1890 to the Second World War, can be argued to have been based upon innovations in electricity, chemicals, synthetic materials and the internal combustion engine:

> In these "long cycles", which average approximately fifty years, the early years are characterised by new technologies being developed and diffused by smaller enterprises. It is during that time that small firms become relatively more important in the economy because of their rapid rates of growth associated with the development of the new technology. Only once the cycle becomes more mature do larger enterprises begin to assert their "control" over the economy.

> At this time some of the new firms established in the early
> stage of the cycle have become large, and existing large
> firms will have acquired many of the small firms having ac-
> cess to the new technology. During this phase of the cycle,
> therefore, large firms become relatively more important
> (Storey, 1994).

The fourth long wave, which began in the 1940s, is now said to be
ending, and to have been based upon electricity, mass production,
petrochemicals and the motor industry. Which means that the
fifth long wave is now beginning:

> . . . and appears likely to be based upon innovations in
> microelectronics, information and communication tech-
> nologies and the like. . . (Burrows, 1991).

David Storey, consistently the most readable and penetrating observer of the small business scene in the UK, agrees with Bur-
rows, but adds a warning note:

> What is being seen today is the upsurge in new and small
> firms whose presence is based upon the "new" commodity of
> information. Explanations of this type, of course, suggest
> that the shift towards small-scale enterprises is merely
> temporary until larger enterprises once again reassert their
> control.

> One of the most striking illustrations of this trend is the
> major growth in numbers of firms in the business and in-
> formation services sector. Not only are existing enterprises
> increasing their demand for externally provided business
> services, but consumers are also shifting their preferences
> towards services, at least partly at the expense of manufac-
> tured goods. Thus it is argued that, since services are more
> likely to be provided by smaller-scale enterprises, the mod-
> ern economy is more likely to be characterised by smaller-
> scale enterprises than was the case in the past . . . (Storey,
> 1994).

This then is one supply-side explanation for the re-emergence of
small-scale enterprise. But there are others, such as cost advantages and fragmentation. Storey's survey of the field identifies
three types of fragmentation:

i Decentralisation of production in which large plants are broken up but retained under the same ownership, by hiving off into smaller plants or by creating new subsidiary companies.

ii A detachment where large firms cease directly to own units but retain revenue links with them (i.e. licensing or franchising).

iii Disintegration of production and innovation: large firms cease to own units of production or innovation but retain control through market power (especially in the case of vertical integration), or, latently, through the power to re-purchase the units *(ibid.)*.

Of course, there may be a simpler reason:

The relative increase in the importance of small-scale enterprises in the manufacturing sector in fact reflects the poor performance of large firms, rather than the good performance of the small. . . . The relative importance of small firms in the United Kingdom as a source of manufacturing employment stems primarily from the declining importance of large firms. During the 1971-82 period, manufacturing employment in firms with under 100 employees remained broadly constant, but contracted sharply for enterprises with 500 or more employees. Thus the relative increase in the importance of small firms reflects the relatively poor performance of large firms in terms of job-shedding *(ibid.)*.

The variety of available hypotheses explains why one consistent factor in the British small business sector over the past two decades has been its use as a political football, with all the inevitable lack of direction that such attention invariably involves. For "political" does not necessarily mean the existence of a policy.

The debate has at least been lively, however; and this in part because, despite the much-vaunted increase in employment in small-medium firms in the UK during the 1980s, many European countries continue to show substantially higher figures, while Japan has had over 80 per cent of the workforce in firms with fewer than 300 employees for some time.

> Several commentators draw a link between the high preponderance of small-medium firms in Japan and the level of adaptiveness of the Japanese economy as a whole. Japan's automotive industry in particular is deemed more efficient than Western counterparts because of its greater reliance on outside suppliers. Small-medium firms can compete through "flexible specialisation" and output flexibility. Depending on industry conditions, this can be a source of extra profit or greater survivability in periods of uncertain demand (Hendry, Arthur and Jones, 1995).

Flexible specialisation, therefore, offers the opportunity for smaller businesses to occupy a central position in the economy, notably within industrial districts in which a number of these small firms form a network that is able to make use of a pool of skilled labour and common facilities, especially in the areas of technology and IT.

> There is a substantial variation and heterogeneity of competitiveness and economic vitality as well as social standards among small firms, both within and across national economies. Thus, there are sweat-shops as well as highly flexible, stable, innovative and independent categories of small firms, often with polyvalent workforces, good pay and extensive autonomy for the worker. The business strategy of these firms is often based on product quality or differentiated products, or on flexible specialisation. It normally requires a skilled workforce and well developed occupational labour markets. The small firms or communities of small firms with good economic and social performance suggest that there is, in terms of competitive strategy, a real alternative to the low cost/low productivity/poor social standard configuration in which many small firms find themselves (Loveman and Sengenberger, 1991).

This idea of "communities of small firms" is an important one; with the firm's attitude and approach to networking being the decisive factor in its success or failure.

> A key to understanding the wide variance in small firm performance and development lies in their "competitive strategy", notably in their links to other firms or institutions.

> Due to their limited economic, financial, personnel and po-
> litical resources, small firms, acting alone, are rarely in a
> position to pursue the strategic behaviour often employed
> by big companies, and therefore they require some sort of
> supportive structure that allows them to compensate for
> their lack of resources. Basically there are three ways of
> overcoming this shortcoming: (1) special protection, privi-
> leges or support transferred to them by the state or some
> other public authority; (2) a foster relationship with a large
> enterprise, or an intermediary organisation (such as a
> bank, university, etc.) which provide various types of re-
> source transfers; (3) creating a community of small firms
> which, through collective self-organisation and co-
> operation, may compensate for the weakness endemic to
> individual small firms *(ibid.)*.

Whichever supportive structure is chosen, therefore, crucially in-
fluences the place that each firm occupies (and freedom of action
that it enjoys) in its local economy, and thus in its industry and
market. In addition, the nature of this structure has an influence
on the ways in which new businesses are created and survive:

> New firm formation is likely to be highest in regions and lo-
> calities with the following characteristics: an industrial
> structure which is biased towards small, independent or
> autonomous units of employment; employees who are en-
> gaged in problem-solving and have customer contact, and so
> possess technical and market knowledge; a concentration of
> technically-progressive large firms; a high awareness of
> past entrepreneurial actions; banks and other financial in-
> stitutions that are sympathetic to the needs of small busi-
> nesses; sources of help and advice available; an affluent
> population; and a social climate that favours individualism
> (Mason, 1991).

But beyond this supportive infrastructure lie the networking
strategies of individual businesses, which need to begin with an
understanding of the potential meanings of the term.

> One of the problems of clearly conceptualising networks is
> their perceived inherent "fuzziness". The network is often
> contrasted with the firm which is seen as a bounded entity,

that is, as having a clear discontinuity in social relations between those who are members of the firm and those who are not. In other words, those who make up the firm and its relations are easily seen as the same by any observer. A network, on the other hand, it is argued, is indeterminate and, in principle, can ramify almost indefinitely. Moreover, networks can be centred on any individual to produce a different network to that of any individual. There are, therefore, as many networks as there are individuals who make up the social system in which they participate . . . (Blackburn, Curran and Jarvis, 1990).

For Blackburn, Curran and Jarvis, there are three main types of network: the first are exchange networks, consisting of those companies with which the firm has commercial dealings; the second are communication networks, consisting of the individuals and organisations with whom the firm has developed relations based upon the exchange of information; the third are social networks, consisting of the family and social circle of everyone within the firm, where these have an impact on the firm and its development.

Network participation by the small business owner may be seen on a continuum. At one pole is compulsory participation in some kinds of external relations. All businesses must have exchanges with the environment in order to gather resources and market their outputs. Networking relations composed of suppliers and customers are therefore at some minimal level involuntary. Where the firm employs othe 3, relations with the labour market also become compulsory.

Further along the continuum participation becomes increasingly voluntary. Thus having relations with the Inland Revenue or a bank are voluntary in a strict sense though in practice for most small businesses they are not a matter of choice. An active marketing policy seeking out further customers is much more clearly voluntary though most small firms will engage in such behaviour even if only at minimal level or intermittently such as when an existing customer is lost. Joining a trade association is further along the continuum and tends to be associated with the size of the enterprise and the kind of economic activities in which it is engaged. Such memberships need to be placed on the continuum depending on whether they are nominal or active: a nominal membership involves less resources and time than an active one. Further across the continuum still and approaching the opposite pole, are activities such as joining local chamber of commerce, small business associations and local political groupings where the latter is motivated by a belief that this will be in the interest of the business *(ibid.)*.

Much of the networking which small-medium businesses undertake is thus involuntary, and must not therefore automatically be seen as a sign of proactive behaviour. Equally, highly proactive managers sometimes have little time to network because of the demands which their increasingly complex business makes upon them.

The forces which influence the ability of a business to network within its environment are thus both internal and external. However, as the essay which concludes Part 4 will demonstrate, learning organisations tend to be actively involved in co-operative learning networks.

# PART FOUR

# INNOVATION IS A STATE OF MIND

# Allegory — Episode Three

*To continue the allegory which was broken off on page 54.*

*To continue the allegory which was broken off on page 54.*

**Third and Final Episode**
Will the third episode of the tale see a resolution of the problems described in the preceding episode?

The alert reader will recall that the transformation of the fishing business affected every part of it: production, because of the new technical equipment; the sales department, due to the introduction of an information retrieval system; the office, where a whole new range of jobs became necessary. But the workforce had not been able to keep up, and so the new directors were faced with a dispute that had brought the business to standstill.

The discussion with the striking workforce is an angry one. The employees are aggressive and filled with bitterness, and accuse the management of failing to anticipate the problems, and of trying to force through change. They demand to be paid overtime, with a bonus for the work overload. The three sons react in different ways to this torrent of criticism. The eldest thinks that the new recruits include a number of trouble-makers who have managed to lead the older employees astray. The youngest is still in a state of shock, and remains speechless because he cannot comprehend how all this could happen to such loyal employees whom he knows well and has always trusted. The middle son, however, is not so surprised: he had seen it all coming when he had noticed how disorganised everything had become, and he had observed the growing dissatisfaction of the workforce as he walked through the factory. Before they had launched into the modernisation, he had suggested that they try putting his father's ideas into practice, but he had come up against the refusal of his elder brother. On the day that the strike is declared, he fills bitter: he feels that all these troubles could have been avoided had they proceeded differently. He speaks his mind quite openly, which doesn't exactly reduce the tension between the three brothers.

The whole community suffers because the business has stopped work, since it is the main employer locally. After a week, at the request of the inhabitants, the mayor meets the three sons and offers to mediate. Meanwhile the employees work with their trades union to come up with the basis of a negotiated solution.

Two weeks into the strike, and with no end in sight, the eldest son accepts that the middle son should try to find a compromise. With patience and tact, he gets both sides talking. They talk during the day, and he works on a solution that will be acceptable to all. After ten days or so, agreement is reached. Work resumes, but the compromise which has been agreed disrupts the whole business, since it had been necessary to deal with all the issues at the same time.

They begin with the reorganisation of the fishing activity, using statistical analysis to compare market demand and the volume of fish caught in the course of the previous weeks. This enables them to calculate the optimal spread of production which should ensure a more regular supply. This will enable the owners, when they are informed of the state of the market, to make the necessary adjustments to their fishing plans. The accountant has worked out how much needs to be turned over for the fleet to be profitable. The work schedule has been stabilised, and the sailors are happy with a £50 monthly bonus.

Meanwhile, in the factory, Paul (who is the new director) meets with the workers in small groups to listen to their complaints and suggestions. It is clear that the best solution will be to reorganise the whole process. But how should they go about it? They decide that the first step should be to go and visit the technology transfer information centre on the mainland. An expert then comes to study the layout of work and the production processes on site. She then facilitates a discussion on her findings and proposals with the management and the workforce.

The employees are organised into multi-skilled teams, each of which is responsible for managing a particular order; quality control is carried out by the team and also at the level of the business as a whole. One member of each team checks that the delivery matches the order and that the packaging is up to standard, before completing the time-sheet for that particular order. In this way, the team's productivity can be monitored. Within the delivery service, another person takes charge of the crates as they are handed over and is responsible for refusing any that he feels will undermine his own target: to reduce to zero the number of client returns. This figure is the way in which the business measures its quality objectives.

Meanwhile, back in the offices, chaos rules. Nobody knows what they are meant to be doing any more, and everyone is blaming the new computer system. Paul meets with the employees service by service and listens to their views. The source of their unhappiness soon becomes apparent: the employees feel incapable of using this new equipment because they do not know how to operate it and they cannot understand where it fits in. They don't know how to interpret the data coming from the market, and so they send the wrong orders and thus cause customer returns. When his turn comes, the director describes how the market now works and explains the thinking behind the new equipment and arrangements. In the light of their experience of the recent crisis, a majority of the employees agree that they need to change the ways in which they work. The few remaining sceptics are gradually won over by peer pressure.

They decide to organise training sessions on how to use the new computer system, which will take place during work hours. Private sessions are even scheduled to ensure that everybody understands the basics. To improve communication (both personal and of information) a glass corridor is added to the building which houses the information system. Three newly recruited young employees ask to be trained to use this system, so as to be able to liaise with the orders service. This will allow them to assist the specialist, and even to replace him in an emergency.

So the crisis ends with what (irony of ironies!) should have been the starting point of the whole process: a diagnostic study of "where are we now?"

It is obvious now that the training was inadequate, that the workforce received no explanation of the strategy which lay behind the new system, that the market information system had been added without anyone considering how to integrate it into the existing services, and so the transplant was rejected. Finally, apart from the three directors, no operator at any level had been consulted over adapting the computer system to the needs and working methods of the employees. So the software produced reams of useless data, while it failed to generate the key information which was required to manage the factory.

Three months have now passed, and a normal working rhythm has returned to every part of the business. Tension has evaporated, and employees and directors are learning to trust each other again. But it was a near thing! They only just managed to hang on to a number of key customers, and defaulting on paying for the new equipment was only averted by the family putting in the money. But the business, although it has found the path again, is not yet out of the woods: there will be a substantial debt hanging round its neck for a good few years yet.

*Chapter 13*

# TECHNOLOGY WATCH

*Contents*

13.1  Technology Watch

13.2  Local Watch

13.3  Using Outsiders to do the Watching

*Two factors today demand that a business should monitor its environment with particular attention:*

- *The changeable nature of markets*

- *The accelerating pace of technical change.*

*But keeping watch implies information and information management (in other words circulating and keeping control of information) with regard to:*

- *The development of customer demand*

- *The possibilities that are opened up by the available technologies at any given moment.*

*And, seen more laterally:*

- *Changes in the strategic behaviour and tactics of competitors*

- *The possibilities for change within the business, as it is forced somewhat beyond its current limits.*

*It would be an error to generalise too much from it, but an example of technology watch in practice may be useful here: Japanese businesses have a voracious appetite for information, but it is their use of this mass of information which is more significant. The Japanese abroad give a superficial but nevertheless accurate image of what is almost a cult of knowledge with regard to everything that goes on outside their own market, and the uses to which they apply this knowledge inside their own business community. The Japanese genius is not simply to be found in the ability to create totally new products, but particularly in the ways in which they continuously adapt what their neighbours are doing. To adapt, one must first understand and only then adopt. Information is crucial in both these phases. The Japanese print one daily paper for every two people (this is twice as many as do the Europeans); Japanese businesses behave in the same way. Jacques Morin says that a company like Mitsui receives 40,000 telexes from across the world each day and that Japanese businesses devote 1.5 per cent of their turnover to information.*

*However, the cult of information is not limited only to its collection; one must know how to process it, to circulate and share it. Indeed, the accumulation of information is also in itself a risk. Although it is not inevitable, this accumulation can end up becoming a mass of conflicting messages out of which it is more and more difficult to extract the relevant information at the right moment. Too much information is potentially as harmful as too little is certain to be.*

*This chapter will only address the question of technological information, while the following chapters will consider the formalities involved in obtaining patents and in protecting the know-how of the business. But the first task is to define technology watch (as opposed to competitor or market watch), all of which exist to define formal systems for monitoring specific aspects of the environment. This chapter relates technological watch to an external Diagnosis and provides details of appropriate organisational features which this involves, in terms of areas to watch and the budgets to attribute.*

**Figure 13.1: Techr. ology Watch**

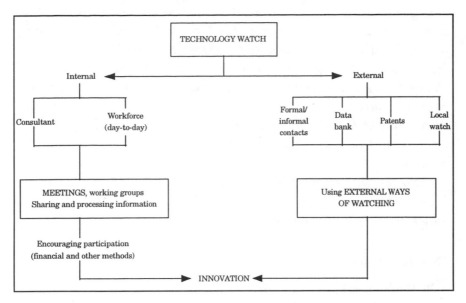

Part 2 of this book (Chapters 3 to 10) dealt with the introduction of major innovations in the business. Placed end to end, these periods of intense innovation are potentially decisive, but they do not cover the whole existence of a business, however receptive it may be to change. Between two major transformations of its (commercial, productive, financial or social) position, the business still needs to make the most of its situation from day to day, attempting to carry out a maximum to improve its competitiveness. The issues needing to be resolved may be framed in a different way and the decisions may not be of the same order, but the stakes for the future of the business are just as high.

These phases between major innovations can be seen as periods for "stabilisation". But they are necessary if a business is to gain control over the disruptions caused by the adoption of an important innovation. This is true for any business, which starts off by paying for change in terms of a very substantial learning cost, before it begins to be able to improve its competitive position. This also applies with regard to the individuals who make up the firm, whose roles or tasks are perhaps undermined by the innovation but who are nevertheless obliged to adapt to it.

*These periods of consolidation occurring between dramatic changes in the business cannot however be viewed as permanent situations. They are rather a kind of embedding zone during which the disruptions linked to the latest innovation are gradually ironed out. Although they can help prepare the next level of change, they must above all contribute to making the previous change a successful one, through projecting onto the market and embedding within the business a new way of doing things, made up of a multitude of permanent small changes in every area of the business.*

*This fourth section of the book, which covers Chapters 13 to 16, is primarily devoted to this reflex of anticipating needs and change which is so characteristic of the innovative business.*

## 13.1 Technology Watch

The possibilities of innovation which a managing director perceives define to some extent the limits of change within the business. This view of what are potential developments can never be defined objectively, since each individual has a personal interpretation of what is involved. In fact several managers in the same firm will invariably have very different judgements on the extent and nature of possible change. The gaps are due to different training, and to individual experience of success and failure, as well as to the self-interest of those concerned.

This potential lack of agreement as to the way forward is also the result of the particular knowledge of each manager concerning every technical question or the organisation in general, as well as their access to knowledge which has been acquired elsewhere, and thus to information gathered on the subject under discussion. This is where monitoring in general becomes technology watch in particular; that is to say the discovery, acquisition, exchange and processing of information.

This task of watching can be fulfilled in a variety of situations: during trips, at trade fairs, in the course of seminars or symposia, while reading, or through consulting a data bank. Unfortunately, these opportunities to be open to the environment are too often seen exclusively from the personal point of view and are regarded as the individual's sole and exclusive territory, in which rights and private rewards have been acquired. On the contrary, however, they need to be systematically exploited by the business as a whole in terms of making a shared strategy happen.

**Figure 13.2: Why Watch?**

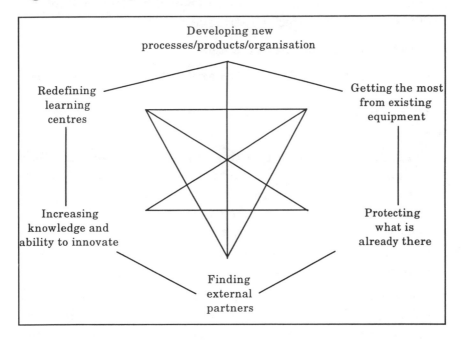

The prime objective of watching the environment is to ensure an early detection of external developments which could have an impact on the business. But, beyond this general goal, watching refreshes the different departments within the business. In particular, it enables learning periods to be shortened, stimulates the internal capacity for innovation and makes a more effective use of the available physical and human capacity. The list below brings out a spiral of causes and effects which can shape the vigilance of any business.

In it are visible the gathering and processing of information linked to technology, to science and to organisation. It will be clear from this that technological activity alone does not cover all the options which the business needs to watch, and must in particular be co-ordinated with:

- **Competitor watch**, focused on monitoring current and also potential competitors

- **Market watch**, focused on knowing the customers, analysing the development of their needs, as well as their solvency; but it is also essential to keep a close watch on suppliers, and notably of their products, prices and strategies

- Watch over **organisational issues**

- Watch over the **social, political and cultural environment.**

More specifically, technology watch covers the following four areas:

1. **Technical information on the components and materials which make up the products of the business.** This consists of the elements currently used by the business and of those which could become substitutes as new materials are developed, technical characteristics improve, or prices come down. New products also appear, but the essence of the new is to be found in the design.

2. **Technical information on production processes.** This involves the equipment used and has less to do with entirely new processes than with new ways of assembling partly independent elements.

3. **Data on information systems** covers a large range of microelectronic systems: computers, robots, software, signal processing equipment (analogical or digital), without forgetting the intangible elements such as programming languages and the connections to databases.

4. Businesses with a significant high-tech content should also collect **scientific information** derived mostly from applied research but also (although more rarely) from fundamental research.

But how accurately can a business define either the precise area to watch or an accurate likely budget for this activity? The areas over which the information specialists in the business will have to exercise their technology watch will need to be defined as an extension of the Diagnosis of the business (see Chapter 3). The specific fields to watch as a matter of priority can be decided as a result of the strategic choice that follows the Diagnosis (see Chapter 5). The picture which the Diagnosis assembles and the results of the technology watch over time come together to form the image which the business has of its environment. It is worth noting, however, that this is a contradictory image: during the Diagnosis, the camera is turned on the business, as it evaluates the internal strengths and weaknesses compared with the competition. On the contrary, during the period of technology watch, the camera is directed outside the business towards the external innovations of competitors, suppliers, and research centres.

The question of what budget to allocate is a much more difficult one, and relevant indicators are hard to come by. Three guidelines should however be given particular attention:

- Information is both limited and expensive

- The existence of a minimum threshold for the watch, beneath which one has every chance of missing critical information and thus threatening the survival of the business

- The current ease with which information circulates within the business and the use which is made of it.

Watching consists of managing a scarce resource (time or money) with reference to an almost limitless amount of raw material (information), on a scale which is right for the business, but within the limits of its ability to absorb and learn. It is very much a matter of needing to navigate between two reefs:

- Wanting to know everything about everything. This leads to the individuals and the business drowning in information, which is not only very costly in time and money, but also creates a situation that nobody in the business will be able to classify, verify, prioritise and synthesise.

- Chasing stubbornly after sources which are equally accessible to the competition, which simply results in one having a share of information which has lost its relevance.

Keeping watch on the environment has a quantitative and a qualitative dimension, and this is as true for a business as it is for a nation or an organism. Volumes as well as links need to be managed, and it is therefore as important at the outset to check the reliability of information gathered, the quality of its processing and the coherence of the diffusion mechanism within the business as it is to check the links between different pieces of knowledge.

At the same time, the real value of the watch process resides in the behaviour it stimulates and the desire to anticipate change which it demonstrates. The finance allocated to this activity cannot alone define its impact, since the means to mobilise are primarily human. One factor which is crucial to the effectiveness of the watch is the choice of the person responsible for managing information within the business, and in particular for transmitting the relevant information to the people concerned. The same is true for the way in which information is processed.

Naturally, the financial and effective human cost will depend to a large extent on the strategy followed, on the sector of activity, and on the size of the business. In a medium-sized firm of 300 people, the responsibility for carrying out the technology watch process could perhaps be given full-time to an engineer, who reports to senior management. For a small firm with around 50 people, it would probably be preferable to use an external agency, and it will be the responsibility of either the managing director or the Consultation Group to manage this agency.

## 13.2  Local Watch

The surveillance network can be divided into two circuits. The first, which is the object of this section, includes the information directly gathered by the staff of the business. The business is thus considered as a productive agent and a consumer of information. The second area corresponds to the use of information that has already been partially processed outside the business by specialists (see the following section). As a result of this the business can be considered to be a consumer of "watch services".

All the members of staff who have any external contact are *information agents*. Their contacts with customers, suppliers, partners, sub-contractors, professional organisations and, of course, competitors are all valuable vantage points, so long as they are exploited. These contacts are established at every level of the hierarchy, in the course of contract negotiations, deliveries or receipt of equipment, informal meetings at trade fairs and conferences, or even within the framework of reflection or information sessions. Each employee thus acquires a new responsibility which is linked to their primary function, and also enriches it. The role of watcher becomes a specific, if secondary, objective.

Some of the information that is gathered in this way results in contractual agreements (catalogues, notices, descriptions, user's manuals linked to the sale of a product or to the transfer of technology). A second area of this information is carried by catalogues (products, prices, trade fair advertising), the professional press or directories. A final source of so-called local information is the direct contacts made on the occasion of open days, organised visits and from direct customer-supplier contact and maintenance offers on competing products. The common-sense rule for exploiting these sources is to begin by getting hold of everything which is accessible, to submit it to a first quick scan, making sure one extracts the most useful information at once, so as not to let piles of documents accumulate. Practising document-piling may impress, but more often than not it reflects dependence on information rather than being in control.

Local monitoring does not stop with the gathering of information: the nature of direct contacts always ensures that relations between individuals are significant, if unpredictable. But although there may be no magic formula, it is worth restating two principles concerning these valued exchanges of information.

Firstly, not everything can be put in a file; nor can one refuse to give anything away. The practice of business secrets still runs strong but has developed considerably, alongside the role now taken by information. Whereas secrets have a primarily defensive character, confidentiality, used advisedly, can become a powerful tool for strengthening a group's cohesion. But half-information can also be destructive and become a source of ill feeling. It is thus important to learn to share knowledge as well as to protect a secret.

Secondly, the strategies for "watching" as well as "being watched" need to be considered together. In any society, the business which watches must accept that it too will be watched. At the individual level, the effectiveness of the observer is linked, over time, to regular co-operation with her source of information and thus to the existence of a reciprocal or shared interest. The information economy is an economy of interdependent systems. In contrast with the "gathering" approach ("I'll get what I can without giving anything in exchange"), its effectiveness is based upon exchange, that is to say from some form of quality contact, involving setting up a "win-win" situation.

Here are three typical situations:

- **Watching a competitor.** A business sends all kinds of signals, if only by its presence in the market. A business which did not communicate would die with its genius still undiscovered. But communication implies a rational transfer of information towards competitors. Businesses operating in the same sector are very much the first source of information for one another. Watching happens here partly in the open, perhaps through professional bodies. But it also happens without the observed business knowing that this is going on, although it is accepted that the latter is conscious of this risk, that it probably does the same in reverse, and that it may be transmitting misleading information to confuse the competition.

- **Communicating with suppliers and subcontractors.** Customer relations and the relationship of reciprocal dependence which they create can help obtain valuable information. The status of client, subcontractor or supplier can enable a business to obtain a considerable volume of information from its partners. The resulting intelligence is in this case more likely to be used for the mutual (rather than conflicting) interest of the partners. This is a major source of industrial innovation.

- **Sharing with one's partners at the beginning of a project.** The launch of operations between businesses, in particular in new and risky areas, can be the occasion for a substantial exchange of information. These situations are as promising as they are risky. Legal protection is important, although it needs to be remembered that too legalistic an approach at the outset can kill the spirit of exchange.

In conclusion, it is necessary to emphasise that the effectiveness of what has been called local watch depends first of all on the spirit in which it is conducted. The watchers need to possess the capacity to be surprised. By developing contacts with the outside world (partnership networks, professional bodies, visits to trade fairs, training, etc.), the business multiplies the opportunities to gather information. These exchanges reveal a degree of readiness in the face of change; they often provide — paradoxically — an indicator of the capacity of the business to protect its own secrets, which it has identified as the core of its own know-how.

## 13.3 Using Outsiders to Do the Watching

The information in which the technology watchers are most interested comes from outside the business. From an initial transmitter, the information is passed along a transmission chain, corresponding to copies or processing, which see the information transformed and integrated into an enlarged or altered framework. Then it enters the orbit of the business, where it is at least as likely to be ignored as it is to be exploited.

The first stages of the circuit travelled by a piece of information are considered as primary. The raw information is contained in the original complete basic documents — periodicals, books, theses, patents. Then this information is analysed (secondary phase), indexed, condensed in less-specialised publications, summaries, catalogues, databases and data banks. Finally, the information reaches a more developed state through authentication, cross-checking, control, translation or synthesis. The exponential growth in the production, transmission and accumulation of information — amplified by the microelectronic revolution — necessitated a substantial formalisation of their collection. Because of this, the information which the business can gather through its own local activity is no longer sufficient to ensure an adequate knowledge of the environment. Once again, the business must strike a balance between its own skills and external support. (Incidentally, the discovery of this alternative path between the market and co-operation won the British economist Ronald Coase a Nobel Prize in 1991). Upstream from the strategic decision, which remains the private domain of the directors of the business, the different phases of information management can be entrusted to outside agents. This delegation of the task of watching can happen at different levels, depending on the human resources which are committed, the budgets which are allocated, and the quality of the information services which the business enjoys. Beyond this issue, the question of the fine-tuning of the receiver also needs to be considered. A trap can be set by a competitor who may be sending counter-signals so as to trigger unnecessary research and expenditure. As in any other field, the performance levels which the provider of the equipment claims may be inflated.

### Trade Fairs, Trade Missions, Study Trips, Symposia . . .

These events provide opportunities for meetings during which it is relatively easy to obtain information from specialists on research in progress or on new areas. (Lists of upcoming industrial and scientific events in each country can invariably be obtained from chambers of commerce or from the commercial section of embassies.)

However, participation in such events requires greater preparation than it usually receives. A real mission plan should be established in advance to identify the types of opportunity to seek out and exploit and the kinds of results which can be expected. During the mission, several steps (including daily meetings and assessments) should be taken. But this is very much an area in which it is wise to take expert advice.

### Data Banks

These information supports have increased in number due to the computer and telecommunications sectors moving closer together. Data banks now exist in numerous fields, so that the challenge lies essentially in locating the relevant data bank. It is not advisable to wander into this minefield without having carried out a thorough preparation to define the field of research. This will certainly reduce the time, and expense, involved in connecting, and particularly in processing.

### The Specialist Agencies within Public and Professional Organisations

There is no shortage of public agencies devoted to supporting innovation, and these offer to accompany businesses on the steep paths of the innovation process. These specialist networks are often to be found halfway between the market orientation of business and the administrative culture of the organisations which support them. This hybrid status is still disconcerting for many managers, even when they do not have a deep-seated reluctance to receive assistance. But there is no shortage of information services which add real value to a business. Test them out.

### Using Consultants

Using external consultants to carry out the technology watch enables the small firm to have consistent access to networks and expertise whose potential bears no relation to their own capacity. The general question of the best way to use consultants will be covered more fully in Chapter 15, which deals with the subject of intellectual property, a legal area for which the use of patent law firms is absolutely essential.

# Chapter 14

# ACQUIRING TECHNOLOGICAL INNOVATION

---

Contents

14.1  The different ways of acquiring technological innovation

14.2  The criteria upon which a decision should be based

---

*A business must go on increasing its capital by continuously intro-*
*ducing innovations into its products, its processes and its organi-*
*sation. The best response to this pressure combines searching for*
*internal solutions and keeping up relations with the outside world,*
*in particular with the producers of science and technology. This is*
*what all large firms attempt to do. Small firms are faced with the*
*same need, but only in exceptional cases do they have the means to*
*carry out research internally. So they must have recourse to other*
*sources of innovation. In most situations, the trump card which a*
*small firm has is neither its own potential for technological devel-*
*opment nor that of its partners, but rather its capacity to transfer,*
*adapt and develop new ways to move forward. The key factor is its*
*aptitude to assemble, more rapidly than those around it, composite*
*solutions which contain a partial innovation, so as to meet effective*
*demand in a market niche.*

*How to get hold of innovation is thus decisive. This chapter will*
*address two issues: the criteria by which a business chooses the*
*right external innovation and the question of how to access it. The*
*following chapter will look at the question of intellectual property*
*and patents, before turning to the way in which the existing struc-*
*tures should be used to ensure a better quality of access to outside*

*knowledge. But the issue here is primarily technological forms of innovation.*

*What are the possible ways in which one can get hold of technological innovation? They are listed below according to the type of collaboration they call for:*

- *To buy or order components or equipment from a catalogue*

- *To buy or have made to order*

- *To make for somebody else or to subcontract*

- *To imitate or to acquire a license*

- *To adapt or add value to ideas existing in other industrial sectors*

- *To be inspired from, to imitate and improve (reverse engineering)*

- *To develop with or to commission research under contract and in collaboration.*

*There is no absolute criterion for choosing the right innovation at the right moment. The history of technology and technological change is littered with the famous failures of brilliant discoveries which arrived too soon but without an appropriate technological context or which did not succeed in creating a new demand. Leonardo da Vinci's flying machines never flew; Denis Papin died in poverty despite having established the principle of the steam engine and having constructed a paddle steamer. More recently, the invention of the car on legs (the Terrastar in 1966) was never brought to production; and the same fate befell the aerotrain, abandoned in the 1980s (but now being reconsidered).*

*The criteria to guide one's choice concerning the most pertinent innovations are at the same time too numerous and of very limited value. The first condition is to have a sufficiently broad understanding of the available options (this was discussed in Chapter 13 on technology watch). But at some point the decision must be made, based upon a limited amount of knowledge. The first criterion here is the cost, notably the cost of access to innovation: cost of the patent, for instance, and of its subsequent implementation set against the anticipated financial results. A reasonable and realistic evaluation of the return on investment is the most basic criterion. This indicator has the advantage of being simple to apply, fast to establish, relatively representative for innovations considered in isolation, and centred around the key variable of return.*

## 14.1 The Different Ways of Acquiring Technological Innovation

The market is indifferent to the source and pedigree of a technical solution. What matters to the buyer are the services from which benefit can be obtained, the cost and the performance of the machines or products acquired, rather than the technological track along which the innovation has developed. The quality of the final product is what counts. However, the quality of the result is always due to a complex linkage of borrowings, acquisitions, improvements, and subcontract manufacturing.

The key ingredients of a good idea do not all need to be invented. There exists at any moment a stock of more or less available and transferable technologies and know-how. But what are these technologies, where are they, to what extent are they useful to the strategic plan of the business, and on what financial terms can the business have access to them?

As has been seen, innovations come in many forms. Some require a long process of research and development, while others can be perfected rapidly, following only a slight adaptation from closely-related applications. Others are available "off the shelf", as is the case when the offer made by suppliers of machines or materials is well adapted. In certain extreme cases, the business may simply take responsibility for the distribution, in the markets it currently covers, of a new product that is already sold elsewhere, within the framework of manufacturing or marketing agreements. This is the solution adopted by French businesses specialising in industrial automation, who preferred to be the representatives of Japanese robots rather than develop their own machines.

The following section describes the main routes to technology, moving from the simplest to the most complicated.

### Buying Equipment or Components from a Catalogue

The industry of equipment goods has a long history, and has reached levels of quality, price and competitiveness which ensure that a very broad choice is always available. It is therefore possible, and much more easily than in the past, to acquire a tool which meets perfectly a specific need of the business, in terms of power, capacity, performance or bulk. (In fact, in many cases, the hardest part is for the business to decide exactly what it needs.) The same is true for the supply of components. Using a catalogue is now clearly the most diversified, the least expensive and the most flexible route to technology.

The skills required of the user are mainly limited to being able to use the tool and to a minimum of maintenance know-how. This is the case for standard machine tools and for computer software to be used in managing stock, orders or accounts.

Buying through a catalogue allows one to benefit from mass-production prices, given that the technology in question is widely distributed, and therefore pretty commonplace. These catalogue purchases also enable the business to fine-tune its watch over its environment because it can gain a firmer grasp both of the assets and of the problems of competitors who are using the same components or the same equipment. On the other hand, buying from a catalogue guarantees no advance over competitors in technological terms, since all competing businesses are likely to find themselves with similar equipment at the same time. Lastly, such a strategy can be reconciled with the pursuit of a differentiation strategy (see Chapter 5), since a business which invests in standard production equipment is able to develop product innovations. In this case the competitive advantage does not result from the mode of fabrication but from specific characteristics (technical, for instance) of the product that it is used to manufacture.

### Made to Order

It is clear that, on occasions, all the machines and components on the market will fail to suit the specific, current needs of a business. When this happens, there remains the option of asking a supplier to manufacture a product to order. Negotiations follow, to determine the specifications required, and the technological, commercial and financial terms. If they are skilfully handled, these consultations between the customer (the potential user) and the holders of the technology in question (the potential supplier) can help one to get quite a precise idea of the state of the art in the field, both in terms of cost, requirements, performance and limitations. The information thus gathered (and possibly multiplied by putting several suppliers in competition) is valuable in measuring precisely the size of the step which the business wishes to and can make. Where the supplier does not enjoy a market position of monopoly or quasi-monopoly, the customer is entitled to be demanding with regard both to performance and timescales. This transfer of responsibility relieves the customer of a large part of responsibility, leaving him free to concentrate on drawing up a precise specification, in the knowledge that the attention to detail that it contains will be a decisive success factor.

### *To Make for Somebody Else or to Subcontract*
Subcontracting has many advantages as a route to new technology, for it enables a business to accumulate skills in a new area of activity by benefiting from previous experience, without having to carry the cost of starting from scratch.

The giver of the order provides his specifications, his route sheet and his work instructions. This may well impose strong constraints, but it can also lead to there being a real transfer of know-how to the subcontractor. The latter will be able to invest in more efficient equipment and adapt its technical production norms, its work rhythms and its timescales. In this way, the subcontracting business is able to progress gradually towards extending and improving its own range.

Two principles can guide a business which chooses the subcontracting route to improving its technological capital.

The first concerns the attention which must be paid to the joint management of risk and the final responsibility of the giver of the order. Car manufacturers, who thought at one time that they could let their subcontractors take the strain by handing over the experimentation and process development phases, have largely reverted, due to an absence of significant benefits, and abandoned these transfers of risk which at the end of the day brought them nothing.

The second principle concerns the length of each agreement. Once again, this is linked to the product life cycle. The technological subcontracting agreement is then closer to industrial and commercial reality than were it to be chopped into yearly chunks. This reflection on the likely length of a life cycle also enables the business to plan the programming of the replacement of the new product by a subsequent generation of products. It is thus a way of managing simultaneously both the innovation and its obsolescence.

### To Imitate or to Acquire a Licence

A licence agreement requires a minimum of compatibility between the contractors on the technical, industrial, commercial, financial and strategic levels. It can represent a relationship that is limited to exploitation of a patent against the payment of a fee. It can also constitute the basis of a broader and deeper collaboration resulting for example in the creation of a joint venture.

The negotiation of the transfer and take-up of licences is a complicated matter, in which it is often necessary (and probably advisable) to call upon a specialist. Among the main points to be determined are the object, the price, the length of time, and the territory where a licence is applicable, as well as the area of technical support. Beyond this initial batch of questions, three clauses call for particular attention. They concern what will happen to improvements, reasons for termination, and the "no competition" obligation. The following chapter will deal with the question of using specialised organisations for dealing with the strictly legal aspects of the negotiation and in particular the study of the validity of patents and their freedom of exploitation.

### To Adapt or Add Value to Innovations Existing in Other Industrial Sectors

It is quite possible for a technology to be unknown in a particular industrial sector, while being of proven value in others. For example, the use of glue is common in businesses which work with wood, but it is new in the cycle industry and largely ignored in the textile industry. A business in either of these two sectors which adapts a gluing process to its own needs therefore gains an advantage over its competitors, even if it has invented absolutely nothing.

This movement of innovations from sector to sector creates multiple opportunities for a small firm. Its understanding of the specific needs of the market that it covers can enable it profitably to transpose (and often at low cost, even when license take-up fees have to be paid) technological break-throughs realised elsewhere. The generic areas of micro-electronics and new materials are examples of almost bottomless tanks of innovation. Such a scenario of sector-to-sector adaptation clearly relies on a wide-ranging technology watch, allowing the detection of information outside of, or in margin of, the natural field of vision of a business. The potential for cross-fertilisation is immeasurable.

### To be Inspired from, to Imitate, to Improve

Copying the ideas of others is all part of the rules of a competitive economy. But outside the framework of the transfer of licences, imitation is restricted through legislation on industrial protection. Counterfeiting, or more generally pirating, is however widely practised. Its development has been accelerated by the opening of frontiers, by the absence or weakness of legal procedures for international protection, by the lack of vigilance on the part of people filing patents, and by the difficulties involved in pursuing offenders.

But the causes are not limited to legal issues. Innovations are concerned with increasingly complex areas in which the organisation methods of different technologies are of greater significance than strictly technical characteristics which are frequently commonplace. Even when specific, an arrangement of components is harder to protect efficiently than the components themselves. This always leaves room for the industrial espionage which occurs naturally and is developing across the globe.

At a more general level, there is nothing to prevent the user of an innovation from understanding its principle and then making modifications, in what is known as reverse engineering.

This practice implies an approach which consists not in rediscovering what has been invented elsewhere, but in seeking to understand the principles upon which it is based. The openness of mind which this demands is often more problematic: it is based on the capacity to detect good ideas from outside and get them accepted inside. This practice brings a double benefit. The cost is minimal (since the research costs were carried by the competitors) and it allows a much faster reaction than is the case when developing something oneself.

### *To Develop or Commission Research under Contract or in Collaboration*

Focused R&D is the most difficult route to innovation for small firms: it is hungry for money, for people, for time, and makes large demands in terms of results and scale. As such, this solution must be seen very much as a last resort. Internal R&D can, however, be envisaged in two specific cases:

- When the strategic option which the business has selected requires the complete appropriation of a state-of-the-art technology and the exclusive use of its applications

- Where all the other routes to a new technology have been exhausted or used.

In most cases, a medium-sized business does not have the means to develop this internal R&D. To get hold of an innovation, it will need to call upon a "producer" of technology such as a university, a technical centre or a research organisation which is able to take charge of all or a part of the development process. This step obviously involves the moving of technological development outside the organisation. Another solution for the small firm can be to take part in a collective development project, side-by-side with partners whose skills complement its own.

Research centres are reservoirs of grey matter which specialise in the production of R&D studies. They put their teams at their clients' disposal to work on innovation projects and bill them for the corresponding outputs. Calling upon a research firm is a flexible solution which allows the business to cross an important technological threshold without having to carry the permanent cost of a development team. This solution also offers financial advantages in so far as the customer pays for using tools which have already been developed, whereas internal research would have needed to develop these tools at the outset. Involvement in this kind of collaboration is generally progressive. Collaboration is divided into separate stages, so as to evaluate accurately the cost and timescales of each one.

## 14.2 The Criteria upon Which a Decision Should Be Based

The technological characteristics of the offer which a business promotes are not sufficient on their own to define the competitive advantages which that offer contains. It is therefore necessary to define additional criteria in order to evaluate the benefits associated with a given technology. This evaluation will allow the business to concentrate on the most appropriate areas to pursue and to make an informed choice of the route to innovation that is best suited to its needs, its means and the competitive situation in which its offer is positioned. Three criteria need to be considered here: the cost, the extent of suitability and the level of exclusivity.

### The Cost

The investment involved in introducing an innovation covers *material* costs as well as the costs of *knowledge* and *know-how*. The intangible component of technological investment does however make a precise total costing difficult. The gross purchase price of a technology (the licence cost in particular) only represents a small part of the final cost. How does one evaluate, in particular, the wealth of expertise compared with resources which are depreciable? In the face of the problem of evaluating the cost of an innovation, most managers prefer to protect themselves from being caught out by an overrun in production costs, which usually results from an overestimation of expected performance, or an underestimation of hidden costs, unexpected costs, the costs of marginal changes, or of the new skills required to put the innovation into practice.

### The Extent of Suitability

The extent of suitability of a technology refers to the capacity of the business which holds the technology to use it, develop it, and find new applications for it. It increases in relation to the extent to which the means, knowledge and know-how are shared.

The different levels of suitability of a technology can be loosely classified in increasing order. They are **almost nil** when the business buys equipment and standard components from a catalogue; and they are **weak** when it orders specific components. When the business has knowledge but little know-how, it enjoys a **theoretical suitability**. Lastly, if it masters the equipment it has acquired, it can be said to have attained a **high level of suitability**.

In other words, when the business has succeeded in adapting the conception of its products to the possibilities of a technology that it is using, the extent of its suitability can be considered as **average**. It becomes **superior** as soon as the business has succeeded in adapting the technology to new applications. It becomes **complete** when the business masters the technology completely, including its development phase.

### The Different Levels on the Exclusivity Scale

To be the sole user of a technology is to be in a theoretical situation of absolute monopoly. Exclusivity as defined by the patents held by the business confers a competitive advantage which is both limited in time (improvements will be necessary) and is only partial (the monopoly may concern, for instance, only a detail in the process or a specific geographical area).

Levels of exclusivity can also be classified on a rising scale. Exclusivity is **nil** in the case of a commonplace technology. It is **limited to improvements** when particular applications are introduced. It becomes **shared** in the context of limited agreements between businesses. It can be **strong** when it is shared by several businesses, in the context of formal agreements. Lastly, the exclusivity is **total** if only one business masters a given technology.

Whichever option is chosen, caution calls, in all cases, for the following:

- A gradual process of enriching the know-how in the business

- Ensuring that the decision and its implementation are paralleled by a combination of training and recruitment

- Early consideration of foreseeable consequences and of the steps to be taken in case of interruption of transfers or a suspension of the supply of the components of the innovation.

In conclusion, it needs to be recalled that the expense of internal R&D is to be used only as a last resort, and then only if one is certain that the question has not already been resolved by someone else and if all other potential solutions have been exhausted. Most of the time it is preferable to go ahead with transfers of skills. In the words of Thomas Durand, the control of technologies through external routes represents a "lever of excellence" to fill the deficiency between the potential of available technological resources and those which are required to develop the chosen areas of activity. However, internal R&D has the enormous advantage of creating, within the business, an intellectual aptitude which helps stimulate the understanding, importation, adoption and transformation of ideas which come from the outside. And it can lead to the occasional brilliant discovery. But the second way of getting hold of technology can never be successful if the first does not exist and operate efficiently. In this context, it is important to guard against the "not invented here" syndrome which is widespread not only in large companies, but also in high-tech small businesses.

*Chapter 15*

# INTELLECTUAL PROPERTY

*Contents*

15.1  Intellectual Property as a Strategic Barrier

15.2  The Acquisition of External Innovation

15.3  Protecting an Internal Innovation

15.4  The Management of Patents

15.5  Protecting a Competitive Advantage

*The rapid change which is taking place in most markets has an impact on the timescale in which technological investments are expected to break even, and so the period in which "innovation outlay" must be recovered is correspondingly reduced. Patent rights enable a business to recover the immaterial rights involved by filing a trademark, an industrial design, or a patent. These can all be seen as part of the assets of a business. Patent rights are laid down and protected by law. However, each case of legal protection involves a declaration being made, and therefore results in the effective publication of something which may well have been considered a manufacturing secret. Any decision to apply for a patent is therefore both irreversible and strategic.*

*The patent policy of any business contains an external element and an internal element.*

*The external element involves monitoring the environment (see Chapter 13) and extends to sending out signals into this environment. It consists of the following:*

- *The systematic monitoring and eventual acquisition of patents and trademarks filed by competitors*

- *The assessment of competitor behaviour in response to the patents filed by the business*

- *The detailed analysis of the supply of imitations or of counterfeits.*

*The internal element aims to gain maximum benefit from innovations created within the business. It consists of the following sequence:*

- *Detecting*

- *Evaluating*

- *Protecting*

- *Being able to exploit innovations, beyond the use to which the business puts them, through granting licences.*

*The final section of this chapter outlines how the management of patents should be organised in the business:*

- *Experience has shown that the whole process begins at the point at which an individual or the management team as a whole take an active interest in these questions.*

- *Using an external expert, usually a patent lawyer, provides the necessary technical dimension to this internal commitment.*

- *This commitment requires quick reflexes to avoid being overtaken by competitors, but it is equally a long-distance race. Its management over time must take account of the fact that the expiry of a patent is inevitable.*

- *The full integration of patent issues into the overall strategy of the business requires above all the commitment of human resources, and therefore the allocation of real time in individual work schedules.*

## 15.1 Intellectual Property as a Strategic Barrier

The rules which surround patent rights represent a boundary to unfair competition between businesses. These rules take the form of patents, for an invention which is capable of industrial application, and of registered designs, for industrial designs.

The principle is very straightforward. The patent gives the inventor exclusive exploitation rights for a period of 20 years in exchange for disclosing the invention. The patent is thus an encouragement to take risks, since it guarantees the creator of an innovation the conditions to exploit it which provide the opportunity to achieve significant profits as a result of the protection provided.

Protection involves two sorts of expense: the first are administrative costs; and there are also the costs of the compulsory publication which accompanies the patent. The administrative costs are essentially indirect. Large companies all tend to have specialist legal departments which deal with these issues but smaller companies have a more cautious attitude. They frequently consider the steps that have to be taken to file patent rights to be unnecessarily complicated compared with any profit which they might eventually obtain from the protection of their innovations. They have the impression that because they are little-known, they are sheltered from risk, and in any case they will not be able to defend themselves if there is a problem. This reluctance often takes the form of a refusal to become involved in what is necessarily a detailed process: they consider time spent in assembling the required legal documents to be non-productive. In France, for example, only 10 per cent of innovative businesses systematically file patents for their most important inventions.

The second type of expense results from the obligation to publish details of the innovation that the business seeks to defend. This can best be described as the dilemma between protection by law and protection by secrecy. The former puts the information into the public domain but forbids its use; the latter, on the other hand, conceals it in order to protect it. Faced with these alternatives, the attitude of an individual firm is as much a reflection of culture as it is a true choice of strategy. Europeans in general have tended to prefer secrecy where business and technologies are concerned; whereas the Americans are much more open about all this; as for the Japanese, they seem primarily to be interested in benefiting the community as a whole, through the wide diffusion of all forms of information within the business (and in so doing they accept the risk of leaks), at the same time as practising the systematic filing of patents. There are as many Japanese as European patents in the aerospace sector, for example, although the Japanese aerospace industry is still very much in its infancy, and the European is highly visible. Japan concentrates the bulk of its protectionist efforts on those specialist fields in which it intends to keep its world leadership: for this reason, over one third of world patents in electronics and in land transport are Japanese. European businesses however tend to have less focused policies despite the presence of a number of specialist firms within most key sectors.

In terms of the global economy, the obligation to communicate technical details of innovations that a business wishes to protect is a key factor in the diffusion of technological expertise. It represents a vital source of information for everyone in an industry. Industry leaders in all the industrialised economies are currently increasing their internal R&D efforts through the use of external sources of information, and the patents registered by their competitors are the main source of information which they use. In further research, they seek to build upon what is already known. There are considerable differences in behaviour, however, between France and Italy on one side, and Japan, Germany and the United States on the other.

## 15.2 The Acquisition of External Innovation

A key factor in a business knowing and defending its strengths is the way in which it keeps an eye on its environment. This aspect of technology watch, described in Chapter 13, keeps track of innovations being used by suppliers, competitors, customers, subcontractors, etc. This certainly needs to be carried out in a systematic way, using databases, trade fairs, publications and consultant studies. It also needs to make maximum use of personal contacts and of partnerships. Through this process it is possible to define the industrial territory which the different players occupy, and also to be able to extend the territory of one's own business through acquiring licenses and through developing the firm's own patents within currently unoccupied areas.

### The Systematic Monitoring of the Filing of Patents

As well as technology watch, consulting the documentation which relates to patents is also the basis of a preliminary study of whether new ideas are suitable for patenting, as well as of the extension or abandonment of exploitation rights. It certainly makes it possible to reduce the risk of investing in what are effectively closed sectors and to detect opportunities to acquire technologies upstream from their development.

The principal institutions for patents are the World Intellectual Property Organisation based in Geneva, the European Patent Office created in 1973 in The Hague and in Munich, and national offices.

### The Acquisition of Licences

The acquisition of licences to patents can be seen to be the necessary entry point to a particular technology. This was certainly the case, on a world scale, for the microchip patented by Roland Moreno: all producers of cards and card readers wanting to use this invention had to pay a fee until the patent expired in 1995. Acquisition also allows a business to keep in touch with progress being made in a specific field, and offers clear advantages by enabling a company to catch up, compared with the lengthy internal development of a technology. Finally, it can be part of a true partnership between a licenser and a licensee, in those cases where their objectives and methods are compatible.

There are a number of different types of licence arrangements possible. They can be either exclusive or simple (several licensees can share the exploitation of a single innovation).

The key points in a negotiation of licensing rights were described in Chapter 14. But the important thing to add here is that *speed of reaction* is absolutely essential. Identifying a potentially interesting patent requires a firm's internal information systems to be highly alert as well as up-to-date, since any delays in internal reaction time may well lead to the loss of important opportunities. It is not possible to get hold of the licence for a promising innovation if a more rapid competitor has already leapt in and concluded a deal.

## 15.3  Protecting an Internal Innovation

### *Protection: Detecting and Evaluating What Can Be Patented and Filing the Patents*

A policy of industrial protection has a double objective: to safeguard and add value to the creative forces within the business; and to use them as tools to develop the idea of innovation as a state of mind. It is not always easy, for technical and human reasons, to identify patentable innovations. The person with responsibility for patents must educate the rest of the organisation to use them as tools.

The critical areas of this delicate task can be summarised as follows:

- **Detecting** innovations, refinements, existing technical improvements

- **Evaluating** which of these are patentable, which are effectively dependant on internal practices, and which should remain secret

- **Filing patents** and taking the necessary actions

- Finally, ensuring that the **management of the patent portfolio** is part of an overall business development strategy.

The evaluation phase is designed to determine what is patentable. In theory at least, this is very simple: a patent can only apply to an invention. This implies that some kind of inventive or new activity has taken place. "New activity" means that nobody, not even its author, has circulated details of the invention in public before the patent is filed, anywhere in the world, and in any form at all. It is also necessary that the invention be able to be produced industrially; scientific discoveries are not therefore able to be patented; computer software is covered by copyright legislation; and aesthetic creations by trademark legislation.

Common sense suggests that it is best to preserve a high level of secrecy around an innovation whose suitability for patenting has not yet been established. A business which publishes in the press, or puts on show to the public, something new that it has not yet taken steps to protect is unable subsequently to patent its invention.

A business needs to check out any innovative area into which it intends to venture, so as to avoid pouring resources into a dead end that has already been legally secured by a competitor. Taking the trouble to research the extent to which there is freedom to manoeuvre in a specific area can certainly enable the development process to be refocused before it is too late. To establish whether a patent already exists, the appropriate organisation (national, European, international) is able to provide details of the current status of technology, and to provide assurance that patents belonging to other players do not block the route. Precedence is established by any document published prior to the date of the filing of the patent. This can consist of a technical article or of a patent already having been filed. It can also consist of an article published by the author of the patent being examined or of an earlier patent registered by her. A search for precedence can therefore establish the extent of a firm's freedom to exploit a patent.

---

### *European Patents*

*Existing European patents cover (it is the person registering the patent who makes this choice) one or more of the following countries: Austria, Belgium, Denmark, Ireland, France, Germany, Greece, Italy, Liechtenstein, Luxembourg, Monaco, the Netherlands, Portugal, Spain, Sweden, Switzerland, United Kingdom. Since 1978, a single procedure has been operated by the European Patent Office in The Hague for all aspects to do with the Search Report (the search for precedents), and in Munich for the full ("substantive") examination. German, French and English are the three official languages of the European Patent Office, and any patent request can be made in any one of them. This does not necessarily have to take place in Munich, and can be carried out in a number of cities across the European Union.*

*It should be born in mind that despite this single procedure the European Patent Office does not really deliver a European patent (in the same way that there is no such thing as a "world patent"). The coverage is national, and the patent holder needs to confirm the patent coverage in specific states. What exists therefore, at least for the time being, is a bundle of national patents.*

---

### Ensuring a Return: Granting Licences and Watching for Counterfeits

The protection of innovation obviously gives greater substance to the options for action which a business enjoys. As well as direct exploitation, selling a patent or granting a licence, there is also the existing market.

Granting a licence enables the indirect exploitation of a market in which the business cannot itself hope to have a direct impact. Clearly, it is preferable to bank the fees, however small they may be, than to dream of exporting into a market which one does not have the resources to enter. Some licensees enjoy a greater diffusion potential than their licenser because of their industrial or technical weight, while it is certainly possible for the licenser to benefit from any eventual improvements which the licensee makes.

Analysis of counterfeits is the last area to consider in ensuring that a business makes the most of its innovations, since *patents alone are not enough to protect an innovation*. There are no shortage of examples, in fact, of the opposite proving to be the case. It is necessary to remain vigilant to ensure that the monopoly which the patent can bring is respected and to identify rapidly any violation of patent rights. The only way to detect these counterfeits is to carry out a constant analysis of what is to be found in the market.

Counterfeiting is not limited, however, to the production of a copy of the patented object. It can also involve an importer, a wholesaler, or an individual user. A negligent business can therefore be prosecuted for counterfeiting because it has not looked into the question of the free circulation of a component or of a piece of equipment which a third party has sold to it. A further factor to consider is that a business can, after it has made an improvement of a technical nature (involving either the product or its production), be sentenced to pay damages and interest to an efficient competitor, where the latter, quicker off the mark, has patented the same improvement after copying it.

Some businesses have developed a particular expertise in constructing patent barriers. They go about this by patenting all (or almost all) possible solutions to a particular problem and they weave an almost impenetrable web of potential claims. Their intention in this situation is no longer to establish whether the patented solutions work or are of financial interest: what is important is that the net should cover a large area and be impenetrable. The result of this is that another company produces the product, and then almost inevitably gets caught in the net, and trapped in a lengthy and costly legal dispute. In this situation, it is possible to stop a competitor going ahead with production until a large ransom has been paid in the form of licences which do not necessarily deliver a decisive advantage. In this way a patent develops into patents of variants, and of tiny improvements, until the point is reached at which it has become possible to harass one's competitors for every tiny new refinement. This tactic of erecting barriers is however very expensive, and is thus outside the range of most small firms.

### 15.4 The Management of Patents

Legal protection offers a temporary safeguard, and its length is established in law. In practice however it is somewhat shorter than the law decrees, since it is related to the rate of obsolescence of the innovation which is protected.

### *Involving Top Management*

Intellectual property demands significant and consistent management involvement. The choice of whether or not to develop a patent policy and the different aspects of putting it into practice have clear strategic implications. Giving prominence to assets is identical to sharing knowledge. The talents of the business are primarily to be found in its workforce. Organisation charts, procedures and operating manuals are of little value when compared with the accumulated knowledge of each employee. And each time the business loses one of its associates (in the broadest sense), it also loses a slice (however thin) of its knowledge and of its skill. Filing a patent, on the other hand, enables the business, with a single action, to defend its technological capital, and to store and share this knowledge.

Other methods can be used to achieve the same ends: presentations being given on ongoing research; the pooling of knowledge as part of the induction of new staff; establishing an annual research programme; creating new research teams; publicising each new innovation that the business develops. But what the patent particularly provides is a level of visibility and recognition within the organisation which all employees value and respect.

An *offensive strategy* is required for any development of a new market. What this involves is protecting the R&D from the start and providing adequate protection that covers any new advance by the business. Patents are essential in the case of any venture into what is new territory (e.g. expanding overseas) or a new type of process. Legal protection is about increasing and consolidating the firm's own boundaries, before restricting the progress of the competition.

A *defensive strategy* has a more protective objective, which is to preserve the heart of the firm's know-how through filing key patents, which has the effect of containing the expansion of competitors through patents which bar the way. Patents registered in this way can multiply and cover the broadest possible range of basic principles. Barriers raised in this way will only be effective however if they open the way for industrial exploitation and an ongoing process of renewal.

A *dissuasive strategy* is a more subtle one. It consists of hindering the progress of competitors by surrounding them with mines which take the form of multiple secondary patents (marginal improvements on the competitor's know-how). The problems which these cause a competitor can also take the form of a decoy, since they convey the impression of the dissuading business having a particular skill or interest, without this necessarily ever having to be developed.

The largest element in any patents budget should be that of management effort. The budget is thus mostly human. As an indication, the direct cost of a UK patent is in the range of £1,000-£1,500; that of a foreign patent is between £1,500 and £2,000; and that of a European patent covering five countries about £3,500. The charge for a portfolio including domestic and foreign patents is around £140 per year and per patent. These figures are not particularly high, and they remain an almost insignificant element in the total cost of developing an innovation. But the real cost, taking into account the time involved, is many times higher.

**Figure 15.1: Evaluating the Performance of a Patent**

**Assets**

- Increase in turnover (new products, or increased market share)
- Fees received (on licences granted)
- Fiscal advantages and other public grants
- Brand image
- Internal stimulus to innovation and strategic activities.

**Liabilities**

- Cost of time spent in patent filing and technology watch
- Administrative cost of patenting
- Fees paid out on licences acquired
- Documentation costs
- Cost of awareness-raising and training of staff in this area of activity.

*Using Consultants*

Patenting is a profession; patenting is also a strategy. In cases where the managing director of a small firm has neither the skills nor the time to prepare a solid patent application, it is better to hand it over to a patent consultant, who is able to act as an external legal department. The firm obviously needs however to decide on its own intellectual property strategy, and in which specific areas it wishes to protect its innovations.

Once this basis has been established, the consultant takes responsibility for translating strategic options into legal reality. At the same time, she can manage the portfolio of innovations which are already protected, and liaise with the in-house person with responsibility for patents.

Using patent consultancy does, however, raise questions which go well beyond the issue of intellectual property. It is important to bear in mind that this partnership (just like any other) must follow clear rules which cover objectives, the rules of the game and the timetable. Success depends in part on the right questions being asked, and therefore on the expert being fully and accurately briefed.

## 15.5 Protecting a Competitive Advantage

Figure 15.2 below summarises the legal mechanisms which a business can use to protect what it believes to be its competitive advantage. But it should not be forgotten that there are a number of other mechanisms with which they should always be associated:

- Continuous rapid improvement

- Low operating costs

- Targeted niche marketing

- Product design and quality

- Effective branding

- Maintaining secrecy

- Customer service.

## Figure 15.2: Legal Mechanisms for Protecting Competitive Advantage

**Patents**
Patents are granted to those who can lay claim to a new product or new process, or to an improvement of an existing product or process that was not previously known. A patent provides a monopoly to make, use or sell the innovation for a fixed period of time.

**Registered design**
Registering an industrial design gives its owner sole rights over it for an initial period of five years. This can be renewed for a maximum period of 25 years.

**Copyright**
Copyright gives rights to the creators of original literature, drama, music, art, recordings and computer programmes. These allow the creators to control exploitation. It is granted automatically and requires no formal registration in the UK.

**Registered Trade Marks**
A trade mark is an identification symbol which is used to distinguish one company's products from similar products made by others. It is not necessary to register a trade mark, though this confers a statutory monopoly. Registration lasts initially for seven years though this can be renewed indefinitely.

**Confidentiality agreements**
These provide a mechanism for one company to disclose confidential information to another, while protecting ownership and use. In some cases companies are very reluctant to receive such information for fear that it will contaminate their own development work in the area.

Finally, it should not be forgotten that some methods of protection are more effective than others. Research undertaken by Levin in

1987 across 650 companies in 130 sectors produced the following responses with regard to the protection both of new improved processes and products. The range was from 1 (not at all effective) to 7 (very effective). The most effective method in each category is highlighted.

**Figure 15.3: Methods of Protection of Processes and Products**

| Method of Appropriation | Processes | Products |
|---|---|---|
| (alphabetical order) | (average across whole sample) | |
| Lead time | **5.11** | 5.41 |
| Moving quickly down the learning curve | 5.02 | 5.09 |
| Patents to prevent duplication | 3.52 | 4.33 |
| Patents to secure royalty income | 3.31 | 3.75 |
| Sales or service efforts | 4.55 | **5.59** |
| Secrecy | 4.31 | 3.57 |

These results are a timely reminder that legal mechanisms can usefully, where appropriate, support — *but never replace* — the professional and systematic approach to the innovation process which this book has defined.

# Chapter 16

# PARTNERSHIP

*Contents*

16.1  What is Partnership?

16.2  The Rules of the Game

16.3  Obstacles, Risks and Advantages

16.4  Partners and Agreements

*Innovation is an unforgiving adventure. It requires imagination, money, organisation and perseverance. Organisation is an internal issue, but it also involves the business and its partners. Believing that one can stand alone to fight always ends in failure. It is for this reason that there has been such an increase in the number of associations between businesses of the same size, of links created with multinationals, and, on top of all this, the new initiatives which have been launched by public bodies and other organisations. These are all partnerships, in the broadest sense of the term. These associations are to be found in Japan, but also in Canada, Korea, Taiwan or in European countries: Germany, UK, France and Italy. But the most fertile soil for partnership is paradoxically the country which talks the most about the success of the individual: the United States.*

*But what exactly is meant by the term partnership, beyond the current trend for using and abusing it? The origin of partnership can be traced to the simple division of work, but its recent rediscovery owes more to the transformations that are taking place in the world economy, many of which have focused attention on the quality of production. How can one succeed in new and complex activities while remaining remote from specialists who have the skills or advantages which one lacks? One of them may have a real understanding of the technology, another possesses the capital, a third has the industrial expertise or the business assets, while a fourth enjoys real political influence or patent rights. Co-operation, whether it be industrial, technological or scientific, grows out of market necessity.*

*These associations are temporary. They are limited to specific objectives. Partners are as likely to be company directors in industry as they are laboratory directors, financiers, directors of non-profit organisations, or even people with political influence or so-called ordinary citizens. The number of partners can be limited to two or there may be involvement of a number of interested parties. These arrangements are ad hoc, and made to measure. They may be driven by the need to get together in the pursuit of major contracts (power stations, airports or public works), but these relationships are becoming increasingly medium-sized and local.*

*The innovation partnership is built around a common objective; it presupposes a determination to be successful but also the shared acceptance of the differences between the partners and the risk of failure. That is why this form of partnership needs to be solidly structured. It remains hierarchical and conflictual. Each project requires not only specific partners but also objectives, means, a leader, a distribution of tasks, evaluation tools. These projects are, more than others, based on the need to achieve and to share in success; hence the importance that is attached to evaluation. As well as the purely technological risks and the judgement of the market it is also essential that there is a real understanding between people who all have their own interests to consider. But this needs to be kept in proportion: partnership may be a buzz-word but it is more often uttered than practised. Most industrialists do not engage in partnerships; they mount their own operations.*

## 16.1 What is Partnership?

What is interesting is how a partnership begins. An innovation partnership is built around a common interest and several individual interests.

A medium-sized company cannot possibly attempt to wage war on every front. It tries to benefit from the latest technological discoveries made by research centres, or from being a subcontractor to a major group. In the majority of cases, it is looking for work rather than offering it. It may seek satisfaction through buying technology, as and when required, from those who have it already; but this solution is expensive and has its limitations. It can also assemble both the technology in its possession and the capacity to use it. Where it does this, it substantially increases its chances of success.

---

### *Collaboration between Public and Private Sectors in Germany*

*The Baden-Württemberg region is frequently quoted as an example of partnership activity, for in it are concentrated a number of the forces which drive the creation of new high-technology businesses. The support of public institutions is central to this process. This is delivered in a variety of forms:*

- *An office for economic development and technology parks, which provides useful information to assist in finding available premises at the right price, administrative help, and financial support for scientific institutions*

- *The chamber of commerce and industry, which provides advice, financial aid, legal assistance, contacts with other businesses, data banks*

- *The regional labour agency which provides business development grants, assists in setting up technology transfer contracts and provides advice for specific projects*

- *Government agencies for technology transfer. Professionals offer technological advice and facilitate collaboration with universities and scientific institutes*

- *Programmes for technological research and providing finance*

- *Universities, acting as advisors or as a research pool, take part in the creation of new businesses and monitor technology transfer projects.*

---

Competition is an essential and invigorating aspect of partnership, since competing to achieve a strong position in a particular sector does not contradict the principle of association. It is very much part of the relationship between the business and its environment. A competitor can also become a partner with a specific goal in mind and in a particular context. Partnership does however raise the issue of competition in a new way: the need for a business carefully to evaluate how its own resources and those of its partners are used so as to achieve the best possible outcome. This occurs not only through the exchange of goods but also of resources, information, services and know-how. Establishing collaboration does not mean having to eliminate any disparity in skills levels, or to conceal particular private interests or financial difficulties; but it does require the ability to manage them based on agreed rules. This is why economists view partnership as being half-way between the market and the organisation, between competition and cartel.

This sharing of complementary means of industrial development represents a radical change in the logic which underlies the management of production. That is to say, it involves, giving birth to a "new culture" which can provide real answers to new needs.

Local geography and the local economy are key factors in all this. An area in which a number of businesses, services, universities and research institutes coexist and work in related fields is likely to be a fertile ground for partnership. The origins of partnership are often to be found in a dynamic managing director, but the driving force is as likely to be the leader of the local economic development agency, a village mayor or a retailer. Of course, the less an area is dynamic, the harder it will be to think of new ways to create and develop partnership. In this situation, public organisations (both local and national) need to play a more active role, as has been the case in Southern Italy and in other peripheries of the European Union.

The key concept is that of *co-operation*, which can take four forms:

• Industrial and commercial co-operation involving productive businesses

- Technical and scientific co-operation with agencies, service organisations or specialist advisory bodies

- Financial co-operation with banks and venture capital partners

- Co-operation with administrative bodies (national, regional or local) and social organisations (trade unions and other forms of association).

But partnership, like any form of association which involves the pooling of interests, resources and results, does not happen on its own. An entrepreneur is not naturally disposed to discuss her own projects with potential competitors.

The experience of Prato (see below) underlines the importance of regional associations of entrepreneurs, involving all kinds of contacts between different sectors as much as it involves relationships with local government in economic forms of participation (business services, the creation of consortia). It should already be clear that such a highly developed division of activity within a single area requires an increase in the number of organisations representing the interests of the participants.

---

### The Industrial District Consortia in Prato (Italy)

*Prato is one of the key textile production areas in the world. Its production system is based upon a particular organisation, and upon a spread of labour between a substantial number of small firms each of which specialises in a narrow aspect of the production process. This was reorganised as a result of a severe industrial crisis which involved bankruptcies and substantial levels of redundancy.*

*Prato's success is due to the flexible organisation of its production and a highly-developed division of labour which results in a specialised economy. This raises the question of how to co-ordinate the different participants. The key factor in this case was the speed at which collaboration (between the businesses placing the orders and subcontractors), which went beyond simple market relationships, was developed. There were thus savings in transaction costs which enabled there to be a reduction in market-user costs*

*(through the capacity to co-ordinate and thus reduce conflict be-
tween businesses) and an increase in production flexibility.*

**The Sprint consortium.** *The small firms found it in their interest to create R&D consortia, even in cases where they belong to
very traditional sectors. In the 1980s, an association of public and
private organisations, both local and from further afield, created
Sprint to co-ordinate and stimulate innovation initiatives. Among
Sprint's projects, an information retrieval system called Sprintel
offers a range of different services:*

- *Information and management services*

- *Communications services, which, with the support of the
  Videotel network, allow the exchange of messages between users
  and suppliers of information, and between the different Prato
  businesses, "inpenditore" and "subcontractors", of which there
  are several thousand in the area*

- *Transnational services (Pratel and Viatel): the first organises
  an information bank of subcontracting availability, so as to
  match offer and demand on the textile labour market; the second ensures the exchange of information on the availability of
  different methods of transport of goods.*

*The advantages: for each business, this consortium gives access, at
a low cost, to services which would otherwise be out of reach. These
services become indispensable when an area with so many small
firms begins, as was the case for Prato, to need to improve the information flow between very small players in the local economy.
Among the advantages not directly sought by the participating
businesses, but a consequence of the consortium, is the spread of an
IT culture in the area. This will undoubtedly have a real impact on
subsequent generations of entrepreneurs and employees.*

*The disadvantages: each business must pool the information
which it has traditionally tended to guard jealously. For this
reason, the Pratel project, by which the "end-point" businesses were
meant to receive information on the productive capacity available
among the "sub-contractors", encountered a certain amount of
resistance from businesses preferring not to circulate the
information. And of course no information system can replace*

*direct contact between entrepreneurs, which plays an important role in the dynamism of an industrial district.*

---

## 16.2 The Rules of the Game

The implementation of co-operation can be achieved in a number of different ways, which will be outlined below. The decisive factors in its development need to be:

- Explicit negotiations and contracts, in which the forms of association and collaboration are clearly defined

- Implicit negotiations and contracts, when these forms are defined informally or are dependent on other contracts

- More direct pressure, when the will of a small number prevails over general objectives.

It is important to bear in mind that relationships need to be established between people before they can be established between businesses and institutions, which means that the question of confidence is one of the key factors in the construction of a partnership (which always involves some element of risk).

The development stages for this type of association can be broken down as follows:

- Choice of partners by the initiator of the project

- Establishment of agreements on the basis of respective skills and experience

- Development of an appropriate structure for the management of the project

- Search for funding

- Effective implementation of the project

- Evaluation of the project (using criteria determined in advance) while it is still running and at its conclusion

- Sharing the profit on a pro-rata basis according to input (and not just financially).

Contributions do indeed need to be of different kinds, and to relate to each partner's area of greatest ability:

- Human abilities
- Capital
- Land
- Patents
- Commercial and distribution networks
- Training activity
- Fiscal advantages
- Political help, etc.

Where the identification of partners is carried out with a real attention to detail, it will enable the business to create a network of relationships which will bring it significant benefits, which in turn will contribute to creating the next stage of growth.

---

### *A Variety of Contributions: A Brewery in Seattle*

*Unknown prior to 1975, several small firms in the brewing sector have been created on the west coast of the USA. Today there are 19 of them. Paul Shipman created Redhook Brewery in 1981. He only used capital from his family and the bank, equipment imported from Germany and a few local fiscal incentives. In 1985, he wanted to expand and create a new factory. He explains his situation and his strategy:*

*"We increased our turnover by 40 per cent a year, but we could not apply for venture capital (we were not a high-tech business) or for public credits (which required bureaucratic efforts beyond my capability).*

*The city of Seattle offered me an audit at no charge. They looked at a financial package and presented it to the State authorities. The study began at the end of 1985, funding was obtained early in 1987 and the factory opened in December 1987. We created a new factory on an abandoned industrial site, and we added a public bar for retail sales, as well as a nightclub.*

*The total package was $3 million, made up of:*

- *The building acquired by the city and sold to a private developer who rents it to us fully equipped ($1 million)*

- *Local risk capital, including my own ($1 million)*

- *Local bank loan at 1 per cent above the prime rate with a guarantee from the city ($0.5 million)*

- *Loan over 10 years at 9 per cent guaranteed by the federal state ($0.5 million).*

*I naturally retain managerial control over my business as a whole."*

---

## 16.3 Obstacles, Risks and Advantages

There are not always immediate advantages to be gained from implementing a policy which looks outside the business, and involves investing one's own resources with other partners. The sources of the difficulties which many businesses experience with partnership are as much to be found within the economic, social and political context of an area as they are within internal behaviour: a fear of competition, the habit of working on one's own, a lack of confidence in public administrative bodies, the attitude of trade unions or a lack of interest in this local aspect of economic development.

So how should these difficulties be overcome? Reference has already been made to the different opportunities offered by a dynamic region as opposed to more lethargic areas. Very often, the business must rely on external organisations (for instance associations of entrepreneurs) to obtain services and establish appropriate relationships. A business which adopts a cautious approach invariably loses out in the long run. The object is therefore to maximise the advantages while minimising the risks. There is no cast-iron method for ensuring success, but there are nevertheless a number of recurring themes.

The long term advantages of co-operation are not difficult to identify:

- Lower production costs

- Improved consistency in growth, in addition to new products and new prospects

- Strengthening of competitiveness

- Increased scientific and technological knowledge

- Accumulation of skills and increased training activity

- Launch of high-level projects which would still only be ideas if the business had relied solely on its own resources

- Development of relations with specific actors which can be activated whenever they are needed

- In addition, there is the contribution that the business has made to the dynamism of the region.

Whatever the issues which a business wishes to address through co-operation, there is always a large element of uncertainty around which negotiation will take place. Each participant begins with a particular attitude depending on its own resources, capacities or interests. At one moment or another, it may be driven to take a stand against the other players. It is important that this opposition, however normal it may be, does not compromise the objectives of the group as a whole. It is in this situation that the partnership will be forged and developed, in the acceptance that these contrasts, contradictions and conflicts are all within the rules of the game. The management of conflict is certainly one of the most complex issues which must be addressed, but it is also an aspect which deserves serious consideration *before the launch* of the partnership.

### 16.4 Partners and Agreements

The potential partners of the business can be:

- Directors of other businesses

- Managers from complementary sectors

- Financiers

- Managers of service structures

- Administrators of local or public institutions

- Directors or managers of associations or foundations

- Academics, members of professional institutions

- Anybody else.

The criteria for choosing partners are not always determined by reason or particular requirements. Reference has already been made to a number of potentially decisive factors: geographical proximity, the past relations of the players, market forces or financial performance.

There are four key characteristics to look for in a partnership relationship:

- Compatible interests and shared objectives

- Technical specialisation

- Complementary skills and resources

- Synergy.

---

### Synergy in Partnerships

*The single most important precondition for a partnership to work is that the whole must be more than the sum of the parts. Synergy refers to the additional benefits which can be achieved by two or more partners working together, as compared to their potential achievement in isolation.*

*Synergy arises in particular from the differences between partners, who will tend to have very different:*

- *Skills*

- *Knowledge and experience*

- *Resources*

- *Conceptions of what is possible.*

*All these differences are potential sources of synergy. The catch is that they can each also be a source of conflict.*

*When evaluating synergy it is important to consider the costs as well as the benefits. Very diverse organisations find it hard to collaborate, and partnership always involves staff time and energy to make it work; it may involve a lot of organisational and cultural changes on both sides. All this involves expending resources. In short, not all partnerships which can bring benefits are worth the true costs.*

*But synergy does not necessarily or only mean generating extra revenue; it can mean the achievement of a qualitative objective such as improving the working environment. Should this be the case, it is important to ensure that social as well as financial benefits are included in the evaluation process (from* Economic Development Workbook *(1994).*

---

For many businesses, however, there remains the difficulty of finding the right partner with the need, the interest and the will to collaborate. In many cases a business just does not have a sufficient grasp of the basics (in terms of organisation and information) to be able to envisage an outside relationship which results in a formal co-operation agreement. There may, for instance, be disparities between partners (size of the business, ability to innovate, etc.) or competition in the market which can trigger subsequent problems. In addition, there is always the risk of betrayal in which one partner exploits the other in the areas of management and the use of the results. For these reasons, agreements need to be assessed case by case, depending on their complexity.

The case of the Seattle brewery is significant with regard to the types of partnership that are possible between the public and the private sectors with the aim of developing or revitalising an area. It illustrates a general rule that small firms, despite their initial competitive disadvantages and their financial difficulties, manage to achieve generally good results when they establish partnerships with other forms of economic structure.

### Agreements

A partnership relationship finds its expression in a contract, which is the outcome of earlier agreements. The contract determines the nature and the objectives of the co-operation, the ways in which it will develop and the rules within which it will operate. There needs to be particular clarity concerning the criteria upon which the association is based:

- Shared interests and objectives
- The pooling of resources
- How these resources will be used and organised
- The responsibilities of each partner and the choosing of the leader
- The resolution of any conflict
- The method for sharing the outcomes.

Any agreement only exists within a number of "levels of complexity". These are dependent upon the source and content of the agreements as much as on the number and nature of the partners. In every case, the level of complexity can be defined as one of the following:

- **Negligible,** when there is minimal uncertainty and the agreements are pretty standard: the transfer of productive resources, transfer of material and immaterial resources, know-how. In this case, links between partners are clearly determined and cease as soon as the result is achieved and the balance paid.

- **Low,** when the agreements concern subcontracting (transfer of techniques, from the buyer, and of project data and knowledge) or particular commercial agreements (franchising); or when there are more complex technology transfer agreements. In this case, the partners are linked for a long time, until the profits are shared out.

- **Medium,** for agreements involving development and joint ventures (the production of components, new technologies, services) or the formation of consortia. Where this is the case, the structure of the contract must be significantly more flexible.

- **High,** for the most advanced forms of joint venture, minority or equal share participation in major projects. This level of complexity involves a closer interaction between the partners, with regard to the activity itself and the splitting of the risk.

---

### *Partnership in Practice:*
### *The Risk Capital of the Friulia*

*This example involves the local financial help given to businesses in the small Italian region of Friulia, where the presence of small firms is related to intense activity in the tertiary sector.*

*In this region, as in many others, small firms are not only dominated by large companies from the point of view of turnover, but also with regard to management, fixed investment, and in particular the financial charges which they have to bear.*

*Large businesses find it easier to obtain long-term credit; while small firms have to rely on financing their own growth, which inevitably limits their growth potential. Against this background the Friulia-Veneto region formed an investment company, called the Friulia, which collaborates with the banks, the IRI investment companies and the insurance companies.*

*The role of this company is to promote regional economic development through taking a minority share in existing companies or new companies which have their registered office in the region.*

*The Friulia supplies low-cost risk capital, medium and long-term finance at reduced rates, financial assistance, and technical, organisational and administrative advice. It has also established a number of agencies which offer services to business.*

*The majority of the bank's funds are used in interventions intended to revitalise businesses or specific projects by providing medium- and long-term loans. In its first 20 years of operation (from 1967-87) the company invested the equivalent of £200 million to assist over 230 businesses.*

---

# Readings and Arguments 4

The essay at the end of Part 3 concluded with the argument that learning organisations tend to be actively involved in co-operative learning networks. It is the concept and reality of the learning organisation, characterised by "innovation as a state of mind", that this final essay examines.

Like the preceding essays, however, this one begins with a summary of the key arguments of Part 4.

---

### Key Arguments

- Japanese businesses are hungry for information, but it is the way in which they apply this knowledge to achieve competitive advantage that is more important.

- Technology watch is a formal (internal and external) system for monitoring specific aspects of the environment.

- Technology watch consists of discovering, acquiring, exchanging and processing information.

- Other forms of "watch" with which technology watch needs to be co-ordinated are a watch over the market, competitors, organisational issues and the social/political/cultural environment.

- Watching consists of monitoring a scarce resource (time or money) with reference to an almost limitless amount of raw material (information), on a scale which is right for the business, but within the limits of its ability to absorb and learn.

- Any member of an organisation who has an external contact is an information agent.

---

- Strategies for "watching" and for "being watched" need to be considered together.

- There are a number of possible ways of getting hold of technological innovation. Flexible businesses test them all and evaluate the costs and benefits in terms of clearly defined strategic objectives.

- As in any other area of the business, intellectual property requires a strategy, which is both implemented and evaluated, and is applied inside and outside the business.

- Patents alone are not enough to protect an innovation, or any other competitive advantage.

- Partnerships, in a variety of forms, are at the heart of strategic innovation.

- Any evaluation of partnership needs to take tangible and intangible elements, as well as short-term and long-term costs and benefits, into account.

For all the lip service paid to the free market, the European success stories in the area of regional economic development demonstrate the important role played by support agencies in sustaining effective collaborative networks involving small-medium businesses.

Writing as early as 1961, Burns and Stalker identify that:

> . . . if the greatest invention of the nineteenth century was the invention of the method of invention, the task of the succeeding century has been to organise inventiveness. The difference is not in the nature of invention or of inventors, but the manner in which the context of social institutions is organised for their support (Burns and Stalker, 1961).

The term "industrial district" is a more recent coinage; but the most frequently cited examples (such as the Emilia-Romagna region of Italy) all underline the advantages which accrue as a result of a concentration of related (and often complementary) activities. But its contribution is not solely economic:

One characteristic of the industrial district is that it must be viewed as an economic and social entity, which means that there exist close interrelations between economic, political and social spheres; and that the functioning of one, in this context the economic, is determined by the functioning and the organisation of the others. The success of the districts no longer depends solely on the "kingdom" of the economy. Major social and political factors are also important (Pyke and Sengenberger, 1990).

A key function of the industrial district in terms of helping smaller businesses to overcome one of their natural disadvantages is enabling a pool of specialised labour to be maintained and managed in a way that allows sectoral skills to develop and circulate freely. This improves the job security of individuals; and enables companies to plan for uncertainty with greater confidence:

> To manage by continuous anticipation is to mobilise sufficient information to construct a flexible organisation — one which is not overtaken by changes in the environment. In this sense it is just as necessary to plan ahead in a printing firm with 50 employees as in a large business, because accurate forecasting is even more vital if one depends on a single market. The best way of coping with uncertainty is thus to prepare one's organisation for it, just as the best way to reduce breakdowns is preventive maintenance of the equipment. Anticipation is a sort of preventive maintenance of the organisation (Riboud, 1987).

This form of preventive maintenance is just one of the aspects that characterises the best innovative companies in Japan and elsewhere. It is a mixture of innovation as a system and innovation as a state of mind, and is thus reflected both in their strategy and their structure across the whole business; and in their openness to the environment.

> Innovative SMEs are characterised by their determination to confront and work with technological change. Their organisational structure reflects this determination. The most dynamic enterprises are those in which development policy depends on more than one person (contrary to most family

enterprises) and in which external input is welcome. In an effort to assess the ability of enterprises to absorb external know-how, Meyer-Krahmer drew distinctions between large, medium and small degrees of openness to external input. Enterprises with a large degree of openness always managed to obtain necessary information, regardless of their location, through formal contacts with experts, consultants, research laboratories, and so on. Such was the case, for instance, with technologically sophisticated enterprises whose location was independent of information factors (footloose enterprises). Enterprises with a medium degree of openness showed a preference for solving their problems internally and had only occasional, informal contacts with outside sources of information. However, the local availability of experts, universities, technical schools, and so on increased the likelihood that these enterprises would resort to external input. Enterprises with a small degree of openness were impervious to the local availability of information owing to their inherent reluctance to use external resources (resistance to outside interference) (Maillat, 1988).

Culture (like innovation itself) is a much abused term; and so Schein's precise definition is doubly valuable:

> Organisational culture is the pattern of basic assumptions that a given group has invented, discovered, or developed in learning to cope with its problems of external adaptation and internal integration, and that have worked well enough to be considered valid, and, therefore, to be taught to new members as the correct way to perceive, think and feel in relation to these problems (Schein, 1984).

But the culture for innovation, as described by Handy in his accessible and wide-ranging classic, is not without its difficulties:

> The task culture is job or project-oriented. Its accompanying structure can be best represented as a net, with some of the strands of the net thicker and stronger than the others. Much of the power and influence lies at the interstices of the net, at the knots. The so-called "matrix organisation" is one structural form of the task culture . . .

The whole emphasis of the task culture is on getting the job done. To this end the culture seeks to bring together the appropriate resources, the right people at the right level of the organisation, and to let them get on with it (Handy, 1976).

The dominant culture is thus that of the team:

A benefit sought by companies that employ team structures is improved quality of work life. Most human beings seem to prefer jobs that include social interaction, and work teams provide opportunities for small talk, development of friendships, and empathic reactions from other employees. Such socialisation was anathema to Frederick Taylor, but although there may be a trade-off between productivity and socialisation on the job, it is also true that alienated, unhappy individual workers are no more productive than overly-socialised team workers. Our experience with companies that have adopted team structures has been that productivity has either remained stable or increased. It is well known that team structures are common in Japan, whose companies are noted for their productivity levels; process improvement and quality control in Japan, according to Ishikawa, "is a group activity and cannot be done by individuals. It calls for teamwork" (Davenport, 1993).

The identification of the individual with the objectives of the team and of the organisation, linked to the way in which teams are created and adapted to respond to specific needs, makes the task culture demanding to manage and to participate in. It is especially appropriate, Handy argues, where flexibility and responsiveness to the environment matter:

You will find the task culture where the market is competitive, where the product life is short, where speed of reaction is important. But the task culture finds it hard to produce economies of scale or great depth of expertise . . .

Essentially control is retained by the top management by means of allocation of projects, people and resources. Vital projects are given to good people with no restrictions in time, space or materials. But little day-to-day control can be exerted over the methods of working or the procedures without violating the norms of the culture. These cultures therefore tend to flourish when the climate is agreeable, when the product is all-important and the customer is always right, and when resources are available for all who can justify using them . . .

However, when resources are not available to all who can justify their need for them, when money and people have to be rationed, top management begins to feel the need for control methods as well as results. Alternatively team leaders begin to compete, using political influence, for available resources. In either case, morale in the work-groups declines and the job becomes less satisfying in itself, so that individuals begin to change their psychological contract and reveal their individual objectives. This new state of affairs necessitates rules and procedures or exchange methods of influence, and the use of position or resource power by the managers to get the work done. In short, the task culture tends to change to a role or power culture when resources are limited or the total organisation is unsuccessful. It is a difficult culture to control and inherently unstable by itself (Handy, 1976).

The fragile nature of this "innovation as a state of mind" culture is underlined by Kanter's "Ten Rules for Stifling Innovation":

1. Regard any new idea from below with suspicion — because it's new, and because it's from below.

2. Insist that people who need your approval to act first go through several other levels of management to get their signatures.

3. Ask departments or individuals to challenge and criticise each other's proposals. (That saves you the job of deciding; you just pick the survivor.)

4. Express your criticisms freely, and withhold your praise. (That keeps people on their toes.) Let them know they can be fired at any time.

5. Treat identification of problems as signs of failure, to discourage people from letting you know when something in their area isn't working.

6. Control everything carefully. Make sure people count anything that can be counted, frequently.

7. Make decisions to reorganise or change policies in secret, and spring them on people unexpectedly. (This also keeps people on their toes.)

8. Make sure that requests for information are fully justified, and make sure that it is not given out to managers freely. (You don't want data to fall into the wrong hands.)

9. Assign to lower-level managers, in the name of delegation and participation, responsibility for figuring out how to cut back, lay off, move people around, or otherwise implement threatening decisions you have made. And get them to do it quickly.

10. And above all, never forget that you, the higher-ups, already know everything important about this business (Kanter, 1992).

The vulnerability of the learning organisation (and its capacity to move towards becoming a "frustrating organisation" which Kanter's Ten Rules reveals) cannot be overstated, in the same way that the achievement of learning organisation status can never be assumed to be durable. But this uncertainty does not prevent the best writers and practitioners from enthusiastically characterising the learning disciplines, or from searching for ways to facilitate the process of making them happen. Peter Senge's indispensable *Fifth Discipline Fieldbook* (1994), for instance, buzzes with theories, games and ideas over practical exercises to develop and extend the learning skills of the organisation and the individuals within it.

Underlying this whole approach is the conviction that the ability to learn faster than one's competitors is, in particular for a smaller business, the only sustainable competitive advantage.

The key function of a learning organisation is therefore its capacity to create a context in which people are continually learning how to learn together, and thus extending their capacity to achieve the results they are seeking.

# PART FIVE

# INNOVATION POLICY AND PRACTICE

# Chapter 17

# INNOVATION POLICY AND PRACTICE ACROSS EUROPE

*Although the general principles of science and technology are accessible to anyone who feels inclined to consult books, data banks or other sources of knowledge, those of innovation call for a particular environment, context and atmosphere. A number of chapters in this book have shown that innovation cannot burst forth spontaneously, in the middle of the desert, even if one is above an oil well. After 40 years of success, the best-known industrial districts in the United States, such as Silicon Valley or Route 128 around Boston, have shown that the true creative genius of a business is never as energetic as when working alongside others and being able to enjoy effective communications structures, prestigious universities, and the co-operation of skilled professionals in areas outside their own competence, experience or mission. The same is true of Europe.*

## 17.1  A New Approach to Regional Policy

There has been, over the last decade, a move throughout most of the European Union towards strengthening economic growth at the national level through a series of policies that focus on a regional approach to stimulating competitiveness and the capacity to generate sustainable growth. This approach has been built upon a new awareness of the fact that differences between regions do not only find their expression in the symptoms that are wage rates and unemployment statistics, but that other factors are the true causes: infrastructure; levels of education and training; quality of the environment; availability of capital; skills level in the labour market. Attention is being paid, therefore, to reducing imbalance in the factors which influence the potential for development, rather than to development itself.

The logical next step in this argument is that there are also significant differences within an individual region; hence the need to develop sub-regional and even local policies, and to ensure that suitable structures exist to be able to make a difference at the appropriate level. This is *subsidiarity in practice*.

This obviously has implications in terms of the balance of power between the national and regional/local levels (and this at a time when national governments have also been involved in transferring a segment of their power to the European level). Without exception, however, those policies based upon decentralisation have only been effective where there has been (as in the case of Spain) a real transfer of the power to carry out decision-making and implementation at the regional level.

This has necessitated the creation, or at least the strengthening, of regional and local authorities; and has also led to a new role for small-medium companies. Whereas, in the not so distant past (and in the present, in parts of the UK, for example) a key political objective was to attract multinational companies into an area, there is now a growing awareness that directing resources and attention towards the strengthening of the small-medium sector's ability for learning by interaction is more cost-effective and has a more substantial and sustainable impact.

For this to succeed, there need to be dynamic public and private organisations which identify with and understand the regional economy and labour market. Once again, it is those regions of Europe which have strong chambers of commerce, local banks and venture capital funds, as well as trade fairs and proactive information centres, which lead the way in terms of the measurable impact of their economic development policies.

---

### Measuring the Regional Potential for Innovation

*The European Innovation Monitoring System (EIMS) has carried out a critical analysis of the different methods which are used to measure the innovation potential of a region. Rather than identify a single "best practice" approach, the research highlights six elements of a regional innovation system which need to be taken into account:*

- *Analysis of supply and demand in terms of innovation services*

- *Investigating extent of current networking between businesses*

- *Assessing current R&D activity within companies*

- *Assessing strengths and weaknesses of regional environment*

- *Evaluation of current policies and their impact*

- *Inventory of current level of technological expertise and application.*

*Research by Hassink (1992) on regional innovation policy, while underlining regional differences in context, detects three favourable conditions for effective regional networking:*

- *A similar mentality between the different organisations involved (businesses, universities, support bodies), which can be assisted by exchange of staff and by the encouragement of active membership of chambers of commerce, etc.*

- *Given the increasing complexity of technical problems, technology transfer offices need to specialise in fields of particular local importance, and be able to signpost customers to other appropriate agencies where necessary.*

- *Co-operation needs to be established between the different support agencies in a given area. Gaining and sustaining this partnership approach is a key function of the regional authority, which also has the role of co-ordinating local players to lobby effectively both the national government and the European Commission.*

*Where these conditions are found, and where national government and regional government allocate appropriate resources over a medium-long period, collaboration leading to innovative behaviour is likely to have a positive impact on regional competitiveness.*

---

This new approach to regional policy clearly implies a different role for central government, which needs to retain its interventionist role but to become a co-ordinator and facilitator, leaving the areas of decision-making, implementation and evaluation to the appropriate regional or local level, whatever their political allegiance.

This co-ordinating role has several elements:

- Ensuring good communication (in the broadest sense, including the exchange of new ideas) between the European Union and the regions, and between different member states and the regions.

- Ensuring good communication at the regional level between different bodies and interest groups. This may well include providing the required funding to ensure that a necessary body is created, or to ensure that redundant bodies are either merged or dissolved.

- Involvement in ensuring that the appropriate infrastructure is in place. This covers the areas of "heavy" infrastructure (e.g. rail and road links) as well as "light" (education, training, information networks).

- Ensuring that all local businesses, whatever their size, have access to appropriate information and expert advice (rather than subsidy).

- Ensuring that the conditions are in place to develop a local workforce with appropriate skills to contribute to future growth.

There is thus an intimate link between the development of innovation within small-medium firms, and regional policy as a whole. In both cases, national, regional and local players, be they public or private, all have a role to play.

## 17.2 Support Networks

Relations between a business and its partners stimulate close links which give solidity to the practice of working together. The more complex the nature of the competitive environment, the greater the need for co-operative relations to take over from conflictual ones. These local relationships form networks which can stimulate the increase of wealth of each partner while reducing transaction costs and forming genuinely effective organisational systems. These networks have geographical foundations (local or regional), productive foundations (clients and suppliers), as well as financial and technological foundations.

States, local communities and other public bodies have fully grasped the importance of these networks and the productive energy which inevitably results from these local relationships. All over the world they have led to the creation of a variety of related agencies, such as the *technopole* (Bari, Milan, Turin, in Italy, or Meylan, Nancy-Brabois, Sophia-Antipolis, Toulouse in France are a few among many). A technopole is built around the idea of bringing together businesses which are world-class in areas where they are likely to provide services for one another. For example, a computer programmer who specialises in CAD can take part in an architectural practice which is building an electromagnetism laboratory, which tests the reliability of the components used by the computer manufacturer, whose computers are used by the CAD specialist.

However, not all European cities can become the headquarters of a technopole, and the successful examples listed above should not be allowed to conceal the much greater number of technopoles which never managed to develop. Should there not then be a place for another form of network, midway between the lonely competitive business and the technopole?

This is precisely the role of the ***Innovation Relay centres***. The idea upon which they are based is that businesses need to be able to draw upon a extensive range of highly specialised organisations, and that in return these organisations must have a sufficient number of customers to be economically viable. These organisations are often public or semi-public organisations (like the Training and Enterprise Councils (TECs) in the UK) which recycle state subsidies and share a common objective of revitalising the industry of the region or of a sector, and of offering advice or technical assistance in the field of innovation, the introduction of new technologies, or even management and organisation.

The current trend is towards the measurable improvement of services offered, which are of a more commercial nature than in the past (services are charged for, if only at their marginal cost). The customers of these regional structures are essentially small firms, although, as is the case in Spain and UK, large companies do sometimes become involved in them.

These structures are generally very small (two, three or maybe up to five people), and they mostly play an intermediary role in the development of partnerships between state, region, business and university.

## 17.3 Different National and Regional Approaches to Innovation

This central section of the chapter is devoted to an examination of different national and regional approaches in Europe to the task of facilitating innovation in small-medium firms. It is not intended to provide comprehensive information on current schemes, since any business interested in these will be able to get hold of up-to-date information rapidly using the Internet or their local Euro-Info or Innovation Relay centre. Nor is it suggested here that successful approaches can automatically be reproduced elsewhere, given the differences in business culture and behaviour that one can observe even between two such geographically close cities as Liverpool and Manchester.

The objective is rather to highlight the different strategies which have been adopted, and the potential for networking and learning which this variety represents; since any British company wishing to develop partnership with a Norwegian business, for instance, can only benefit from being aware of the innovation policy (which is a faithful reflection of a less tangible innovation culture) to which the potential partner is exposed.

The seven cases described below cover a range of policies and practices: beginning with the uniform but effective French approach, we move on to consider the different ways in which German regions react to specific issues. This is followed by consideration of a long-established Danish practice, and then two examples of proven excellence from Norway and Ireland. The two final examples are from Southern Europe, from Spain and Italy; and both illustrate some of the benefits that a regional approach to meeting the innovation needs of small-medium firms can bring.

### France — The Seamless Garment

Decentralisation has perhaps been the major change to have taken place in France since 1981; but the initiator and driver remains the state. The first reason for this is simple: tradition cannot be changed solely by legislation. Nor can the behaviour of small-medium businesses, which have continued to develop their traditional networks, which reflect sectoral and professional rather than regional boundaries. The second reason is that the dominant public schemes in the area of innovation are still national in vision.

The overall picture thus remains, for all the talk of regional diversity, the seamless garment beloved of the central planners.

In France, there are two particular sources of innovation support:

- The first stone to turn over in the search for support in the field of technological innovation is to make an application for the available public funds, by contacting the organisations responsible for their distribution. These outlets are numerous. Some work together, while others are completely independent from one another; and some guard jealously their autonomy and compete directly with other institutions. The largest is the

National Agency for the Development of Research (ANVAR), a key element in the French innovation policy, which federates the current public support programmes.

- The second source of support is local networks. The infrastructure in France ensures that there are always several poles of technological excellence at a reasonable distance from any business: the technological wings of public institutions (a university, a technical centre, an active local or regional group) with a specific technical or sectoral orientation.

These sources are described in more detail below:

*Public Programmes*
Although they are the main beneficiaries of public programmes for innovation, small firms still make relatively poor use (in quality and quantity) of what is available. A lack of familiarity with the procedures, administrative red tape, different mindsets, a mistrust of public officials and agencies are all criticisms (sometimes well-founded) that are directed at these organisations, but more often they are exaggerated and insufficient to justify a refusal to attempt to make use of a range of subsidies, reviews or other forms of technical assistance of increasing quality.

The National Agency for the Development of Research (ANVAR) is a key player on two counts. Firstly because it enjoys a considerable presence on the ground (there are 24 regional offices) which puts it in touch with the whole range of actors involved in spreading innovation: consultants, laboratories, other innovating small firms, technical centres; and secondly because it manages a large number of the government-funded initiatives to stimulate innovation.

---

### ANVAR Funding

*In 1990, the agency provided help for innovation to the tune of FF 1.42 billion. A large part of this budget (just over FF 1 billion) is used for business-innovation projects, with 85 per cent of this amount in 1990 having gone to businesses with fewer than 500 employees. This form of intervention in "business-innovation" covers the costs of perfecting a new product or process with a technological content, from the first studies up to the preparation of the industrial and commercial launch. It includes expenditure sub-contracted to the providers of specialised services, including sub-contracting to laboratories or research companies, and internal expenses such as personnel, the purchase of equipment and components, the cost of acquiring licenses and models, and technical trials and demonstrations.*

*The grants made by ANVAR for these business-innovation projects (there were 1,216 successful applications made in 1990, receiving an average grant of just under FF 1 million) are an advance which is paid back if the project is successful.*

*A percentage of the total grant is to be used to strengthen the innovative capacity of an organisation through encouraging the recruitment of researchers. This covers a maximum of 50 per cent of the internal and external expenses linked to the recruitment of a researcher, and resulted in the recruitment of 440 people in 1990. Other forms of involvement include: the development of European partnerships, notably through the EUREKA programme; grants for the creation of innovative businesses (FF 320 million in 1990); technology transfer (FF 80 million in the same year), and specific support (FF 100 million in 1990) with a view to developing research of particular value to small firms.*

---

## Local Forms of Support or Signposting

The simplest solution remains for a business to look for (and to find) the skills that it requires on its doorstep. This proximity may be geographical, as in the case of regional laboratories or the engineering faculties at the local university. Or it may involve turning to those organisations with which the business has already worked, or which have been recommended to it.

**Universities**. Relations between industry and universities tend to be characterised by equal amounts of mutual attraction and incomprehension. Businesses would like to hire intelligent heads, but frequently do so without knowing very well how to take advantage of the reservoir of grey cells which the universities allegedly house. Universities, for their part, would like to see their research applied in a way which adds value both to themselves and to the economy as a whole, but tend to have little confidence in the constraint of immediate profitability when research calls for a much longer-term approach. Beyond this deep cultural divide, the two worlds are no longer as foreign to each other as they were in the past, and collaboration is increasingly visible and effective. Collaboration with universities does enable certain research and development projects to take shape; and one advantage is that the universities have the advantage of highly-qualified researchers and considerably lower costs than elsewhere.

Universities, for instance, benefit from agreements with the National Agency for Research and Technology (ANRT). These involve subsidising a business (up to FF 90,000 or 14,285 ECU per year over two or three years) in exchange for employing a doctoral student to research a thesis which interests both the business and university. The student works within the business and the university continues to supervise her work.

---

### Connecting the University and Industry — DIVERGENT in Compiègne

*Since its creation in 1972 the University of Technology at Compiègne (UTC) has trained engineers in four divisions (mechanical engineering, chemical engineering, computer science, biological engineering). At present, 350 engineering diplomas are awarded each year. During the three-year course, two semesters are devoted to work placements in businesses. This means that the university has 700 industrial contracts per year. This phenomenal level of contact is further reinforced by the presence of a large body of research students (about 30 per cent of the total number of students), whose research is usually the product of a collaboration between the academic and the industrial worlds.*

*Two particular lessons can be drawn from this experience:*

- *Although, over fifteen years, the UTC stimulated the creation of several businesses which in turn created around 200 jobs, many attempts failed through lack of method and finance.*

- *The company known as DIVERGENT came into existence in 1987 alongside a non-profit association, GRADIENT, specialising in managing academic and industrial research contracts. 50 per cent of the company's capital came from GRADIENT, with the other 50 per cent being divided between 44 academics and five businesses which grew out of the university.*

*The aim of DIVERGENT is to insert a study phase between the initial idea for a project and its technical development so as to analyse its economic and technological credibility and increase the chances of success. This results in a development plan which covers the different stages from searching for finance to marketing. This study phase involves an estimated cost of from FF 50,000 to FF 200,000 (from 7,940 to 31,750 ECU).*

*After an initial pre-selection of projects, applications are put to a committee of 17 people, including bankers, industrialists and representatives of the regional council and regional development agency. At this stage, to ensure that a proper distance is maintained, DIVERGENT is intentionally not represented. Each project selected receives a loan from the banking pool so that DIVERGENT can begin the study. In the event of the idea not leading anywhere, the regional council pays back 75 per cent of the bank loan, the banks assuming the remaining 25 per cent themselves. The initiator of the project thus takes no capital risk, although FF 5,000 (795 ECU) are required to register and to pay the interest on the bank loan. In order to develop its links with the exterior, the UTC in 1991 established a Transfer Centre consisting of a meeting space, a business incubator and a technology park.*

**Networks of Technical Centres: The Example of CETIM.** Most professions have put into place mixed-economy structures, known as *technical centres,* charged with gathering and disseminating specific technical and scientific knowledge. Businesses can make use of the resources of these centres to contribute to their technology watch, to negotiate collaboration with their French and foreign partners, and to commission research and development projects. It is perhaps worth giving a few examples which illustrate the range of services on offer to businesses.

CETIM is one of 18 industrial technical centres within the CTI network. It specialises in mechanical industries and is partly financed by a professional tax. The services that it provides can be grouped under five headings: direct aid, supply of software packages, training, technological information and a standardisation role.

To complement these centres, other agencies specialise in innovation support: the national Agency for the Development of Industrial Automation (ADEPA), Regional Centres of Innovation and Technology Transfer (CRITT), Regional Funds to Aid Consultancy (FRAC) and Regional Agencies for Technical and Scientific Information (ARIST), which are briefly described below.

---

### Four French Agencies that Specialise in Innovation Support

### *ADEPA*

*The Agency for the Development of Industrial Automation is an organisation which transfers new technologies to small firms. Its expertise covers the following:*

- *Carrying out industrial automation audits*

- *Assistance in drawing up terms of reference for consultancy projects*

- *Technical back-up in modernisation projects*

- *Providing information and training for decision-makers, technical directors and application technicians.*

*It also manages, for the state, the distribution of financial assistance for the acquisition of high-tech equipment and software.*

### *CRITT*

*There are around 150 CRITT (Regional Centres for Innovation and Technology Transfer) in France today. They stimulate the awareness of small firms with regard to new technologies, help them to develop new processes and products, and contribute to the training of personnel. Launched in 1983, these transfer centres are usually the result of existing teams getting together (universities, engineering schools, laboratories, etc.) to provide a link between the needs of industrialists and the technological skills in the region. The state is also involved, since the Ministry with responsibility for Research funds 50 per cent (as does the region) of the set-up costs of these not-for-profit associations.*

*A few CRITT are generalist, but the majority specialise in a particular technology. They have technicians and consultants who will offer the first hours of their services at no charge, with any further work being charged at the market rate.*

### FRAC
*Since 1984, the FRAC has encouraged businesses to use the services of external consultants. Studies cover many subjects related to the organisation and development of industrial automation functions, industrial design, strategy, the choice and design of equipment, quality management, and even market research. Reserved specifically for businesses with fewer than 500 employees, this fund is provided by the Ministry for Industry and the Regions.*

### ARIST
*Twenty-seven ARIST (Regional Technical and Scientific Information Agency) were created around the regional chambers of commerce and industry. This network offers three kinds of information services: "Arist-flash", "Arist-synthesis", "Arist-Alert". Downstream, the ARIST provide specific advice. The national network of ARIST has an information retrieval system connected to 400 international data banks, a substantial documentary library, and a network of correspondents in France and abroad.*

---

### Germany
The nature of local innovation policies in the German *länder* vary considerably. In some states, such as Bavaria, they are a relatively low priority; in former industrial areas, like North Rhine-Westphalia, they tend to be reactive to economic problems, and to be based upon a sectoral approach; whereas in a region like Baden-Württemberg, which has a more dynamic small-medium firm sector, there are a number of policies aimed at strengthening

the ability to innovate, with technology transfer being of particular importance.

Inward investment thus remains a low priority, with the key objectives (their weight varying from region to region) being the support of small-medium enterprise, and the encouragement of the broad diffusion of technology. In short, what economists like to call "strengthening the endogenous potential".

The key players in this innovation landscape reflect these objectives. Whereas in Britain the regional tier of government is missing, and the chambers of commerce tend to ignore the issues associated with innovation, the main players in German regions are regional government, the chambers of commerce and institutions of higher education, which have learnt to adopt a practical approach. Public-private partnerships, however, remain a low priority.

The service offer reflects the objectives and the support infrastructure, with technology centres being the cornerstone of innovation policy. Baden-Württemberg, for example, has a dense network of 120 Steinbeis technology transfer centres, which receive funding both from federal and regional authorities. In addition, the region has established 10 technopoles close to universities, offering subsidised support to young companies at the cutting edge. It needs to be recalled, however, that the objective of subsidy is viewed as being to increase a firm's capacity to act on its own.

One aspect of the German approach which has received ample evaluation is the *user-friendly simplicity* of the advisory networks. This transparency is frequently contrasted with the plethora of often competing bodies existing within the UK economic development field. But the picture is far from uniformly perfect: the Ruhr area, for instance, lacks an umbrella organisation to co-ordinate innovation initiatives.

But the major weakness of this impressive structure (and this often comes as a surprise to businesses from elsewhere in Europe whose knowledge of the German market is based exclusively on distorted and invariably outdated anecdotal evidence) is the extent to which the past success of many German small-medium businesses has encouraged complacency and a resistance to change, at a time when the environment is increasingly bumpy and fast-moving. Inflexibility and slow reactions characterise the dinosaur rather than the gazelle.

### *Denmark*

One aspect of the Danish approach to innovation is of particular interest here, for it demonstrates the importance of traditional behaviour in even the most innovative circles. For Denmark enjoys a dynamic small-medium sector without having some of the mechanisms of which the Italian "industrial district" is so proud.

What the Danish economy does have, however, is a tradition of training that seems to have its origins in the agricultural sector at the end of the 19th century. What this created is something that has long since extended into other sectors: the creation of personal networks for mutual learning based upon contacts made in the course of training. Successive improvements to the structures delivering different forms of lifelong learning have widened the impact of these networks beyond sectoral boundaries, and led to the creation of the type of relationships between individual small businesses and their regional markets which have the coherence and dynamism of those found, for example, in Japan.

It is a way of stimulating the growth of learning behaviour in individuals and organisations that has much to offer other regions.

### *Norway*

Between 1989 and 1992 the cornerstone of Norwegian innovation support for small-medium firms was the BUNT programme, funded by the Royal Norwegian Council for Scientific and Technical Research. The programme aim was simple: to link the use of new technology with strategic planning; while its execution saw the development of a cohort of consultants who took over 300 industrial production companies through a three stage process consisting of a general strategic analysis (3-5 days) known as the pilot project, more detailed work (15-17 days) on specific aspects of this project, and a final implementation phase which was the responsibility of the company.

The BUNT programme was also linked to other programmes in areas such as technology transfer, the objective being to create a coherent and practical approach to strategic innovation, through which the company becomes a "continuously learning organisation" of the kind that this book has advocated and described.

The same practical approach (linking strategy to implementation and evaluation) is to be found in current programmes such as the SME analysis programme (which aims to strengthen a firm's capacity to make the right decisions in connection with innovation and finance) and the network programme (which stimulates national and international co-operation on procurement, marketing and production).

What these brief Danish and Norwegian examples show is that smaller, peripheral economies are able to overcome these alleged disadvantages through the development of innovation strategies that are ambitious, forward-looking and distinctly practical. Most of all, they are geared to encouraging small-medium firms to look beyond their national boundaries for partnership and markets.

### *Ireland*

The Irish position, both in terms of the size of the domestic market and in geographical situation, is not unlike that of Norway. So it is probably more than a coincidence that Ireland too has developed a National Technology Audit Programme, administered by Forbairt, which looks at how technology can be most effectively used as part of the total business. The process itself is similar to the BUNT scheme; and the objectives, whether they are achieved in all cases or not, are a statement of intent and culture which by their very existence stimulate participating companies to adopt an innovative approach: to keep Irish manufacturers at the forefront of the requisite technology, business organisation and manufacturing systems which are essential to compete in the global marketplace.

These examples of what is known as "best practice" interestingly come from nations which are no larger than many regions. They are the proof that the "think global, act regional" approach is not merely a plot originating in Brussels, Edinburgh or Barcelona to trick national governments into relinquishing power which they consider themselves to have used creatively and successfully in the past.

## Spain

The example of Barcelona is a particularly interesting one; as is that of Spain as a whole, which demonstrates that for a national economy, as for a small firm, the pre-requisite of successful change is acceptance that change is necessary. Decentralisation in Spanish industrial policy was encouraged in 1992 by the industry ministry whose power would, in the short term at least, be diminished by the process. The whole process was an acceptance of the extent to which Spanish industry had changed, and that local factors were essential to successful social and economic development. This encouraged the ministry to redefine its role as a promoter of the potential of local partnerships and systems of production to generate improvements in productivity, and thus of Spain's competitive position.

The decentralised model allows, in theory at least, the flexibility to tailor policy to distinct systems (see below). This means that enterprise policy as a whole, and innovation policy in particular, can be shaped to meet business needs and regional particularities. Catalonia, for example, has a number of programmes which together form a coherent local strategy which would be unthinkable on the other side of the Pyrenees: specific programmes target the development of small co-operatives, the modernisation of the arts and crafts sector, or R&D forecasting for firms with a clearly defined project.

---

### Small Firm Innovation Networks in the Valencia Region

*Entrepreneurial activity in the manufacturing sector in the Valencia region was not capable of keeping up with the rate at which firms were closing down. This factor, linked to the fact that over 70 per cent of the region's GDP comes from small-medium firms, shaped the strategy behind the setting up of IMPIVA, a private institution financed by the regional government. Its goals are the following: encouraging firms to modernise their production structures; introducing new activities into the region to diversify the local industrial base; assisting firms to become more innovative; promoting inter-firm and international co-operation.*

*Research by Javier Ors Valle shows how these objectives are achieved: through setting up a number of Technological Institutes; through establishing BICs; and through subsidising innovative activity within firms (training, new machinery, etc.).*

*The key outcomes of the early years of this approach were the following:*

- *Increasing co-operation between firms stimulated the way in which the Technological Institutes and BICs have improved networking and opened channels of communication.*

- *The newly decentralised technological structure means that firms are now geographically close to firms with whom they may collaborate.*

- *Some of the centres within this technological infrastructure have become highly specialised in specific sectors, and thus become centres of excellence that have stimulated continuous improvement within local firms. This specialisation has also made it easier for firms to find the kind of assistance that they require.*

- *There has been improved collaboration between the different support agencies in the region, from which all parties benefit.*

---

This is an example of a policy *by region*, rather than a policy *for region* as devised by a central government. In Spain, this approach is in its infancy. Whereas in Italy, after over a decade of activity, the impact of a regional approach to innovation is beginning to appear, both in the areas of genuine success and in those of evident failure.

### Italy — The Patchwork Quilt

The Italian network of technical service centres for business only came into being in 1985. It has continued to grow rapidly, but there are still too few cases in which the services on offer are sufficiently advanced to be really effective. Nevertheless, even in such a traditional Italian region as Veneto, a change in attitude among small firms can be perceived. Despite what can be described as being at best self-centred behaviour, and suspicion of any approach that contains an element of association or change, a combination of information, communication, training, the flow of ideas and contact with the experience of others is beginning to bring about changes in behaviour. Entrepreneurs know that they can no longer count on the myth of Italian fantasy and creativity. They increasingly need organisation, co-operation and investment in research and development.

*The Service Offer*
The offer of services to businesses in Italy is generally unstructured, and too often short-lived, highly-specific and carried out from a purely local perspective, shaped and dominated by the needs of a small group of politically-favoured companies. Over the past ten years, however, a part of this offer has become both more stable and more focused. New organisations have developed through the public sector or professional associations. These have shown a particular interest in training, which has tended in the past to exclude small firms through its exclusive focus on the large public companies.

The service offer has a number of characteristics:

1. **An unstructured offer.** Until recently, there was no policy of offering training and consultancy to small firms in a way which could be evaluated in terms of its impact; nor was any effort made to follow businesses beyond the initial consultancy. The offer still remains very inconsistent at the regional level. In the Mezzogiorno, for example, it depends entirely on public sector initiatives. In this case, it has had a demonstrable impact and is often innovative. However, the spread of initiatives is very patchy, only being taken up outside the initial area on rare occasions. The offer is reduced, therefore, to a number of one-offs, instigated by users, trainers, consultants or local politicians with patronage in mind. In the North, on the other hand, the situation is different: some centres (Bologna, for instance) are extremely active, although others have not yet taken any significant steps forward.

2. **A short-lived and somewhat ad hoc offer.** Services are often provided by companies who view this as something of a

sideline, and have no strategy or real pedagogic objective or capacity. They simply follow public funding and initiatives for a while, and then move on to the next when the money and (frequently artificially-stimulated) demand dries up. In contrast with what can be observed in some other countries, this policy has neither brought about a behavioural change nor stimulated a significant demand for training. It must be said that the services offered to small firms in particular tend to be additional activities, and reworkings of old recipes. The training wings of large public businesses (or their suppliers) work in the external market by recycling their packages to medium-sized businesses before targeting the small business market. The result of this is that, where specific objectives exist in the mind of the small firm, they are only occasionally achieved by the provider of the service.

3. **A local offer with a significant political edge.** It is often a chance to repay favours, or for local politicians to extend their power base. This is particularly true in the case of professional training centres in regions with a special status and of services supplied by trades unions. It is only recently, as a result of dramatic changes in financial companies, that a structured offer such as "research-cities" has developed. Carrying out a national, or even a regional inventory of service firms remains a massive task.

*The Institutional Framework*

In the start-up phase of new businesses, the state and the regions offer facilities or financial incentives. Italy has no shortage of legislation in this field, which very much shapes the services which are offered.

**National Legislation.** The 1980s saw a proliferation of initiatives from the state, from associations, and occasionally from large companies. In 1982, specific funding was set aside for technological innovations, but in practice it was only taken up by large companies. A new law of 1986 established "extraordinary measures for furthering and developing activities of young entrepreneurs in the Mezzogiorno" and pushed the regions towards targeting their action according to the size of the business. This law was to stimulate the development of new activities by young entrepreneurs in the South by encouraging new co-operatives and companies created by young people in almost all sectors of the production of goods or services. Still more recently, a law for the development and innovation of small firms (Law 317 of 1991) was to redefine the objectives of public intervention along three major lines:

- The financing of small firms, specifically using risk capital

- The development of real services by consortia, the direct involvement of businesses and partnership between the public and private sectors

- The encouragement of the spread of new technologies by allocating the first tranche for 1991.

**Regional Legislation.** The main sources of support for small firms have been in the hands of the regions since 1977. The state is restricted to providing the legislative framework, to voting the overall amount which is to be devoted to credits and to special cases in which intervention is deemed appropriate. The regions define and implement policies. This autonomy also applies to funding from the European Union.

*Service Centres*

Almost all Italian regions have created *centres for technology transfer,* or make financial contributions to them. There are around 40 such centres, of varying size and range of services. These particularly concern the finance involved in acquiring high-tech equipment and the research and eventual development of new products or productive processes.

But Italy is also remarkable for its *service centres.* These include a complex range of "real services":

- Computer services and data banks

- Technology diffusion

- Test and analysis of materials and prototypes

- Certification and approval

- Training

- Consultancy in organisation and marketing.

The notion of real service also covers a few basic services for the management of public activities (e.g. cultural). It is a way of intervening in the local production cycle, in the way in which strategy is formulated and planned, as well as in the modernisation of businesses.

This section will concentrate on "real services", therefore, without covering financial services, although these were where the service centres began.

**The Regulatory Framework.** The regions and the decentralisation of the powers of the state (which were already set out in the Constitution) were established in 1977. Apart from the regions with an "extraordinary status", characterised by ethnic problems or by geographical isolation, the regions were not given any authority in the area of industry. In theory they must limit themselves to developing the territory, to craft industry and to professional training. But, since 1970, the regions have been successful in creating finance companies which in a few cases have taken the form of economic development agencies. These companies are increasingly the principal source of intervention in industry.

*Service Centres with Their Origins in Finance.* The regions have followed different policies with regard to real services. These can be classified as follows:

- Regional agencies for the technological development of local productive potential. These are non-specialist agencies which have a similar role to a regional financial structure through presenting themselves as the point of reference for the regional policy for innovation, and by taking (in a few cases) initiatives to form other service centres.

- Inter-business service centres crossing sectoral borders are involved in several sectors at the same time.

- Inter-business service centres of a sectoral kind, i.e. offering services for a specific sector.

- Zonal integrated service centres create zones equipped with basic infrastructure and supply the services to stimulate the co-location there of a coherent range of activities. This category includes the research-city consortia, described below.

- Research and application centres were set up exclusively in the Mezzogiorno and specialise in the spreading of new technologies (transformation of agricultural products and computer technology), but they have also developed research activities.

This summary reveals the extent to which activity is fragmented, not only between regions, but also within an individual region. This results in much duplication and wasting of resources. The current evidence would appear to suggest that those centres created in response to a specific sectoral need, and which have broadened their offer and skills to cover new services or new sectors, are the most successful. On the other hand, those centres created initially to supply generic "technological transfer" services have repositioned themselves to offer highly specialised services.

*From Science Parks to Research-City Consortia.* The **science park** is defined as a public initiative whose co-ordination is handed over to universities or public research institutions. There are only two science parks which are operational: Tecnopolis in Bari and the Area di Ricerca in Trieste.

Tecnopolis was created by the Centre for the Study of Advanced Technology (CSATA) and the University of Bari. Born in the early 1970s, it covers 160 hectares of which 1,600 m² are enclosed. In this area of the University research laboratories, training and technology transfer centres were established, along with centres for the research and development of productive activities. There is now also an incubator which can receive fifty young businesses. An information retrieval network links the whole with the key European cities of Milan, Barcelona, Paris and Stuttgart.

The Area di Ricerca in Trieste is managed by a public law consortium, created by national legislation. It was formed in 1975 following the establishment in Trieste of the headquarters of an international physics centre which had the objective of creating a research institution open to the Third World.

The term *technology pole* is used when the roles of university and industry which characterise the science park are reversed. The promoter of the technology pole is private industry, which requires collaboration with the scientific world to carry out applied research. In addition, many large Italian firms now have large chunks of industrial property effectively lying fallow, as a result of the impact of automated production processes, and are taking steps to make the best possible use of them.

The Pirelli-Bicocca project, around Milan, is an application of "innovation transfer". Its aim is to attract scientific research teams and to ensure the transfer of the results to a group of high-tech businesses in the services sector. Facilities for the production of prototypes, process engineering, as well as engineering consultancy and all computer services have been added to the research facilities.

The Tecnocity project is based in the area around Turin, Ivrea and Novara. The promoter is the Agnelli Foundation, which made the Lingotto zone available for the project, with the backing of the CNR which introduced an "information retrieval counter" linked to different data banks. Tecnocity aims to improve communication between existing businesses which have opted for an innovation strategy.

The *research-city consortia* specialise in the promotion and transfer of technology, and owe their existence to a 1982 law which allows for the financing of consortia programmes by the IMI fund for applied scientific research (the IMI is a public institution created in 1931 for the medium and long-term financing of the industrial sector). The city-research consortia are the result of initiatives by the IRI and CNR with support from other public and private institutions, such as universities and chambers of commerce and industry. They are already established in Genoa, Milan, Rome, Catania, Pisa, Padua, Agrital, Naples and Venice, and have three principal objectives:

• High level research through the creation of centres of excellence

- The transfer of technology to small firms
- The development of new models for university training.

They represent a considerable opportunity for the creation and realisation of research and development programmes at a local, European and international level.

---

## An Exceptional Centre: The CUOA

*Alongside the tried and tested creations are to be found all kinds of hybrid arrangements, such as the University Consortium for the Study of Business Organisation (CUOA) which has its headquarters at de Altavilla Vicentina (Vicenza). This organisation brings together the Universities of Padua, Trent, Venice, Verona, and Udine, several public institutions and many businesses which operate in the manufacturing, banking and commercial sectors.*

*Created in 1957 as a postgraduate training institution attached to the engineering faculty at Padua, the CUOA is today one of the most dynamic management schools in Italy. Training takes various forms, including Masters degrees in business organisation, in marketing, in communication and internationalisation, in systems engineering and integrated automation, in project management and technology, in construction engineering and installations, in addition to a substantial continuing professional development programme.*

*The training activity has also involved young entrepreneurs whose projects have received approval from the Committee for the Development of New Activities by Young Entrepreneurs. This group were all offered a course on "entrepreneurial know-how". This kind of training has been delivered on a large scale to top managers in small-medium businesses as well as to the managers in the financial companies which work with them.*

*But this activity is not limited to training. Know-how is also communicated through research activities targeted at the specific needs of businesses. With this end in view, the CUOA has extended its research in three directions:*

- *The management of public or private businesses*

- *Technological innovation and organisational innovation*

- *The new professions (including an observatory of new technologies).*

*In 1991, with innovation in mind, the CUOA established links with the technological park of Atlanpole (at Nancy in France). The objective of this partnership is to deliver technology transfer agreements between the scientific laboratories and local industry in the two areas.*

---

This overview concludes by describing some of the current trends in Italy which continuing European integration is likely to encourage. Despite the enormous variations from region to region, this is a sector which is growing rapidly, while acquiring new professional skills and methods for creating effective synergies. In general terms, there are a number of trends worth noting:

- In the area of public sector initiatives: within a public system in deep crisis (and partly because of this crisis), co-ordinated policies are being developed, in particular at the regional level, in the fields of training and planned innovation. In this area, the development of the legislative or regulatory framework of regional service centres, of research-city consortia and of agreements between businesses and universities is likely to deepen and accelerate.

- In the private sector, movement is likely to occur in the same direction, based upon professional, local and sectoral groupings.

The basis of this process will be the increased demand for consultancy and training on the part of small firms. It has driven professional associations to take responsibility for ensuring action by creating their own companies within specific sectors and notably in the field of technological innovation and quality. In addition, a few large organisations have already carried out original experiments in the provision of training and support services for small firms (distance learning, multimedia, use of electronic simulations, creation of specialised data banks).

These are all trends which need to be intensified, and their effect evaluated, if the Italian innovation networks are to mature and to have a significant long-term role and impact. From the evaluation that has already taken place, it is clear that a few agencies are already highly developed and play an important role. But others, especially in the South, exist only on paper and experience great difficulty in existing at all. This confirms, as has been seen throughout this book, the extent to which innovation and the structures which work to stimulate it can only develop within a relatively complex and effective social and economic system.

## 17.4 The Encouragement of Innovation at the European Level

The European Commission has had a significant role in enabling its different member states to define policies and allocate resources to a number of aspects of regional development.

There has been no shortage of programmes which have played a catalytic role in the development of various forms of co-operation between regions. There has been considerable debate with regard to whether it is more beneficial for strong regions to work together (as is the case in the "Four motors of Europe" project involving, Baden-Württemberg, Catalonia, Lombardy and Rhône-Alpes), or for weak regions to work together; or whether it is preferable for strong regions to work with weak regions, but with a middle region as intermediary.

In the innovation field, diversity has a particular appeal, given the belief that all regions have some particular strengths that others can learn from, if not imitate. This approach is supported by the European Commission's 1995 *Green Paper on Innovation*, which defines an innovation system as being:

> the sum total of firms in an industry, the fabric of economic and social activities in a region. . . . The quality of the educational system, the regulatory, legislative and fiscal framework, the competitive environment and the firm's partners, the legislation on patents and intellectual property, and the public infrastructure for research and innovation support services, are all examples of factors impeding or promoting innovation (European Commission, 1995).

The Green Paper is not short on realism: small-medium firms account for 66 per cent of jobs in the Union, and:

> . . . they are a reservoir for the creation of jobs and a source of diversity in the industrial fabric. At the same time, the weakness of these firms in terms of finance, human resources and commercial contacts are a source of concern *(ibid.)*.

It is to combat these areas of weakness that the Green Paper outlines new approaches to strengthening the capacity of these firms to be open to, and to absorb, appropriate technologies; and also plans to contribute to improving the business support structure, notably in the areas of information and technology dissemination, at the appropriate level.

For the business, of whatever size, which has a clearly-defined strategy built around innovation projects and inter-regional co-operative partnerships, the current policy climate and networking opportunities are exceptional.

*Chapter 18*

# INNOVATION NETWORKS IN THE UK

*Contents*

*This book has described the need for small-medium firms to work towards the characteristics of the so-called "Ideal European Small Business", as outlined at the end of Chapter 11. There is clearly much that an individual business can do to improve its "innovation as a state of mind", and the subsequent systematic implementation and evaluation of a logically developed strategy. Were this not the case, this book would not have needed to be written. But there is an aspect of each of the characteristics of the Ideal European Small Business to which external support, advice or information can make a significant contribution. It is the forms which these different types of outside intervention take in the UK that this chapter considers.*

*It is not by chance that the majority of the examples of "best practice" which have been included in the previous chapters are from Italy, France and Germany. The British have historically*

*been excellent inventors, but, for the last 150 years at least, some-
what less effective as systematic innovators when compared with
their continental neighbours. For a long time this relative weakness in the innovation field was blamed on a combination of the
weather, the British character and competitors not playing by the
rules. But there are recent signs that the encouragement of innovation is beginning to be taken more seriously. Some of the approaches which this chapter will describe operate on a fairly small
scale, but one of the characteristics of an innovative business is to
identify and pursue the external source of improvement which is
best able to meet its current and likely future needs.*

*What follows needs therefore to be seen against a background of
Britain's often myopic approach to business innovation in all its
forms, which has led to British companies being described as
being very good at producing yesterday's products better.*

*Government policy over the past 40 years has been, if little else,
solidly consistent: "the innovation process within firms must remain their responsibility". It is difficult to argue with this if one
reads it in the narrow sense of the stakeholders in any business
being fully justified in running it in any way which they consider
to be appropriate. But the lack of regional infrastructure (as defined in the previous chapter) in the UK ensures that it all becomes
a question of economies of scale; and, in the business world, it is
the smaller businesses which are the main losers when there is
government "laissez faire".*

*The range of potential support which this chapter describes, al-
though it is flattered in some cases to be listed under the title of
"innovation networks", is able to contribute, to a greater or lesser
extent, to the efforts which an individual business or a group of co-
operating businesses make to becoming more effective innovators.
And this infrastructure can make a significant contribution to at
least two of the four things which, according to the 1994 White Paper* Competitiveness — helping businesses to win, *they need to do
if they are to become more innovative:*

- *Change the climate within their organisation to stimulate
  innovation*

- *Develop their people, equipment and processes by training, investing in new technology, and comparing themselves with leading companies (benchmarking)*

- *Take advantage of external skills and know-how*

- *Increase their use, adaptation and development of novel processes and products, including greater collaboration with universities and other research organisations.*

## 18.1 The Innovative Small Firm in the UK

"Made in Europe", the study of best practice in four nations produced by the London Business School in 1993, contains one particularly significant and much-quoted statistic: 67.5 per cent of the British companies interviewed considered themselves to be "completely or mostly world class". But closer examination revealed only 2.3 per cent of them to be world class. This so-called "perception gap" was by no means confined to British companies; but its impact remains. Innovation must be based, above all else, on a realistic assessment and application of a number of key resources:

- Entrepreneurial skills

- Finance

- Human skills

- Plant and equipment

- Potential customers

- Close customer relationships

- Availability of appropriate information for small firms

- Linkages with industry and research institutions

- Economic dynamism of the region.

Where one or more of these resources is lacking, then a company is likely to find innovation more difficult. But, for a business with a genuine desire to innovate, local dynamism should not be seen as a significant barrier, given the key characteristics of innovative firms:

- The development of business contacts and partnerships outside their own area

- The active exploitation of these networks to get hold of new process and product ideas

- The use of this new process and product capability to diversify and export.

Any significant business development begins, however, with the nurturing of innovation as a state of mind. In its pursuit of this, the Investors in People national standard (outlined at the end of Chapter 6) is not intended to be a solution. Managed locally by **Training and Enterprise Councils**, it is a framework which enables a business to structure the closer integration of its "business plan" and "people plan". There is no shortage of supporting programmes to encourage a more thoughtful management culture and attitude to the development of people, but (however expert the change consultant used) it is the people in a business who have the greatest impact on what is the most delicate aspect of "innovation as a state of mind". The issue is so fundamental that it goes well beyond the kinds of solution which any external programme of support can provide. But, as this book has argued throughout, without it the successful management of the innovation process will remain a mirage.

Strategy needs to drive structure. For most businesses, the pursuit of the innovative drivers identified above need to result from and feed into the lengthy and invariably difficult pursuit of improvement within an organisation as a whole. This improvement often implies a radical change in business strategy and behaviour across a broad range of areas (including management culture and attitude to the development of people) which enable a business to identify — and lock into — a supportive environment which does not necessarily have to be local:

- Use of information, forecasting

- Innovative R&D

- Sources of finance

- Development and use of new technology

- Use of external expertise, advice and services.

It is these headings which will shape the rest of this chapter. A number of the programmes and schemes described are domestic, while the rest (and some of the most dynamic) are European. All of these are proof of the efficacy of partnership in practice, and they also serve to demonstrate the value of a number of small units (since national economies and domestic markets are themselves small in some contexts) pooling their resources to achieve not only economies of scale but also a more substantial impact.

## 18.2 Use of Information, Forecasting

Much of the work in forecasting and the gathering of market data is carried out on a European scale, and is more valuable because of it. One of the key barriers to innovation is inevitably a lack of detailed knowledge of the way in which markets are structured and developing. *The European Innovation Monitoring System (EIMS)* seeks to provide policy makers and businesses with information and analysis on the factors which promote and inhibit innovation at company level across Europe. A key aspect of this programme is the *Community Innovation Survey (CIS)* which aims to put innovation and technological change on the same footing (in terms of information) as other important economic issues. At the heart of this project is a database covering up to 50,000 firms, based on a common questionnaire (which is also to be used in the United States and Japan, thus providing global benchmarks). The data gathered will give a detailed picture of innovative activity both within individual companies and at a sectoral level, as well as identifying the key networks within particular sectors.

To many small firms, these projects may appear to be able only indirectly to improve their innovative performance. This may turn out to be the case, but any business or sector should be happy that work is being carried out, with a view to gaining a more accurate picture of what is happening and where things are moving.

What is certainly true is that the most innovative businesses in any sector will be keen to get hold of and analyse the implications of the data gathered.

Within the UK the ***Technology Foresight Programme*** has investigated a number of key sectors (although Figure 18.1 below shows that almost every sector would appear to be key) with a view to assessing the priorities within the sectors, and subsequently recommending the steps that need to be taken at a national and regional level for the strengths of these sectors to be developed and the weaknesses confronted. Once again, the impact on business will be downstream, but it would probably be an error to ignore the findings of a programme attempting to identify the market opportunities which will be the most important to the UK over the next 10 to 20 years.

**Figure 18.1: The Sectors Involved in the Technology Foresight Programme**

- Chemicals
- Communications
- IT and electronics
- Financial services
- Agriculture, natural resources and environment
- Health and life sciences
- Materials
- Retail and distribution
- Leisure and learning
- Defence and aerospace
- Transport
- Food and drink
- Manufacturing, production and business processes
- Energy
- Construction.

Returning to the present, the DTI's *Overseas Technical Information Service (OTIS)* supplies information about technological advances in some overseas countries; and it is possible to subscribe to the service so as to obtain information about specific technologies or markets.

The other important area in which businesses have highly specific information needs is that of intellectual property. The Patent Office, and the European offices referred to in Part 4, offer search and advisory services across this broad (and frequently mine-infested) field.

### 18.3 "Innovative R&D"

The term "innovative R&D" ought to be a tautology. The fact that it needs to be spelled out reinforces the expensive truth that much R&D carried out by firms of all sizes is not new at all, (and so tends to be development rather than research).

For the majority of small firms, R&D should be undertaken within a partnership, and this more for the benefits than for the sharing of costs. *EUREKA* is a European initiative which aims to facilitate market-driven collaborative projects in all sectors of technology, with the objective of introducing new products, processes and services which have a relevance to global markets. The key criteria are listed in the box below. Among the benefits which EUREKA offers which are of particular value to smaller firms are networking and partner search services, and thus the opportunity, whether or not a decision is made to proceed, to meet like-minded companies and to pick their brains (while allowing their potential partners to do the same).

### Figure 18.2: EUREKA: Two Criteria and Nine Categories

> The key criteria for any EUREKA project are that it must comprise at least two organisations from at least two EUREKA countries (European Union, EFTA, Hungary, Turkey, Russia) and that it must involve technical innovation.
>
> Current projects mostly fall within the following categories (some of which are very broad):

- Communication

- Energy

- Environment

- Information Technology

- Lasers

- Medical/biotechnology

- New materials

- Robotics and production automation

- Transport.

Similar to EUREKA in many ways is the European *Co-operation in the Field of Scientific and Technical Research (COST) programme)*. It is also based on collaborative projects, but its concentration on pre-competitive research means that both individuals and research and academic institutions are eligible as well as businesses.

The **LINK** programme is very similar to COST, bringing together companies and research-based organisations to do research which firms can use to produce innovative products, processes, systems and services.

For all of the above schemes, the message from the funding agencies is clear: collaborative R&D brings both immediate and longer-term rewards.

## 18.4 Alternative Sources of Finance

There are a number of grants and other financial incentives available to small firms, and, despite what many businesses believe, it is possible to come away with some money after jumping through the blazing hoops. Nevertheless, neither the chosen strategy of the business nor the relationship between risk and return should be ignored in the search for funding.

### Venture Capital

For many smaller firms wishing to fund innovation projects, venture capital is an option worth considering, not least because it is a way of bringing new specialist expertise and management skills into a business. Unlike banks, venture capitalists are business partners, which is why they are looking for companies with the potential for above-average growth in terms of profit and turnover. A number of other aspects of a venture capital agreement must not be forgotten: the investment is for a limited period; the exit route is defined at the very start; rapid and significant growth and profit are required.

Finding a suitable partner is no easier in this area than in any other. There are however a number of "dating agencies", including venture capital clubs, which draw financial institutions into local networks and into contact with businesses. The clubs are already operating in Scotland and the North West of England, and other areas are now waking up to this approach. For, as the list in the box below shows, venture capital is active in some unexpected sectors (the full list finds the film industry, a notorious financial black hole, in 19th place).

### Figure 18.3: Venture Capitalists' Preferred Industries

1. Engineering
2. Communications
3. Health care
4. Chemicals
5. Leisure
6. Electronics
7. Publishing and education
8. Other services
9. Computer services
10. Computer software
11. High-tech
12. Biotechnology.

### *Grants — National*

Moving on from venture capital to the area of national grants, **SMART** and **SPUR** are probably the two best known UK government schemes in this area. SMART is an annual competition for firms with fewer than 50 employees, which is simply providing funding for a process that every business should be doing annually in the first place: searching for ideas of highly innovative and highly marketable technology. SPUR is a grant (for which larger companies can also apply) towards the development cost of new products or processes which involve a significant technological advance for the industry concerned.

In development areas and in Objective 2 industrial areas of the UK, **Regional Enterprise Grants for Innovation Projects** are available to firms with up to 50 employees wishing to undertake product or process development. But, like a number of the schemes described so far in this section, its title is a misnomer in terms of the definition of innovation which has been given throughout this book, and prevails in most of the European Union. Innovation is still perceived in government (and some business) circles to be a concept that is intimately tied to technology. But this is not in the sense of *technology-pull*, in which technology is summoned to add greater value to the innovation process; this is *technology-push*, in which a company views an investment in new technology as the proof that it is innovative (even in cases where the previous generation of technology was only being used at two-thirds of its full potential).

## 18.5  Development and Use of New Technology

There is no doubt that this section and the following describe a range of benefits which are particularly valued and needed by small firms, notably through the opportunities for partnership and the easy availability of expert advice that a number of the approaches discussed provide. The **Regional Technology Centres**, for instance, although they are somewhat uneven in terms of user-friendliness, are able to offer practical, problem-solving support with regard to technology, training and the whole business. Technology-pull is certainly in evidence. They also have access to a large local network, which enables them to respond rapidly to requests for specialist information or advisors, and are able to offer technology transfer services.

Although it has now been replaced by a programme which involves the dissemination and exploitation of research results, the **Strategic Programme for Innovation and Technology Transfer (SPRINT)** continues to have an influence through its achievements in the area of technology-pull, which substantially improved Europe's ability to innovate and to transfer technology between business sectors and different regions. The importance of this regional dimension is summarised in the box below, but a brief overview of the five areas of SPRINT activity highlights the fields in which a small firm wanting to break down internal and external barriers can receive support. The thinking behind this approach is consistent: innovation itself is cumulative, interactive and complex.

The main areas of the SPRINT inheritance show the programme's emphasis on small firms and their needs:

- **Infrastructure for innovation** — networks of innovation brokers throughout the Member States; specific technology transfer networks for major industrial sectors; training programmes for Science Park managers; strengthening networking between science parks.

---

### The Relative Failure of UK Science Parks

*The first science parks in the UK were set up in the 1970s at Cambridge and Heriot-Watt, but the real growth came during the 1980s, when 36 new parks were set up, with most universities having one. The theory behind science parks was that they would create employment; lead to new firms being established; facilitate links between host academic institutions and park firms; generate "leading edge" firms with a high level of technology.*

*However, as Massey and Wield (1992) have shown, only a small minority of firms on science parks fit this model, and this for three reasons: science parks are property developments, with the owners tending to rent high-status premises rather than selling land; secondly, science parks are not industrial estates: they may be close to academe, but they cut R&D off from physical production; and finally, the simple idea of beginning with research and moving in a linear way towards commercial production is only one of several possible innovation models (there is also "learning by doing" for example).*

*Their contribution to the transformation of local industries by building on local skills and production strengths are thus minimal when set against, for instance, Japanese examples of transforming older shipbuilding areas into centres of advanced marine engineering, or pottery districts into bases for advanced ceramics. The UK science park approach has tended, by contrast, to set "new against old, clean against dirty, sunrise against sunset".*

---

- **Network support** — dissemination of Best Practice in Networking methods; Technology Transfer Days; Investment Fora to facilitate innovation funding.

- **Specific projects** — concentration on existing technological development that requires no further R&D, with a view to identifying and removing barriers to technology transfer (e.g. environmentally-positive technologies; modernising small firms in traditional industries, such as textiles).

- **Innovation management** — recognition of the need to manage the technology; seminars and studies on managing quality and design; managing the integration of new technologies, through a process involving the diagnosis of the way technology is used and of the potential for integrating new technologies and management techniques.

- **Innovation monitoring** — the EIMS scheme, described above, examines ways in which innovation and technology transfer can be improved.

---

### Why the Regional Dimension is so Important for Innovative and Technology-based Small Firms

*Small firms do not usually have sufficient internal resources to integrate new technologies and innovate. Not only do they need easier access to a set of external resources and support services, but they also frequently require an external spur to alert them to available opportunities.*

*These support services can be best provided in close proximity to their target companies, since technology transfer does not necessarily have to be done through licences or through getting hold of research results. The importance of the learning process means that attention needs also to be given to direct personal contacts through training, learning by doing and interaction. For this reason support services need to be developed on a local or regional basis, and to concentrate on providing appropriate and relevant information and advice.*

---

This argument applies equally to the **Value Relay Centres**, which again are part of a European network. The Centres have the dual role of informing companies and research institutions about EC R&D programmes, as well as assisting small firms wishing to participate. Once again, hands-on is the key, linked to the recognition that there is a real need among the small business community for management development sessions that cover the whole range of business activities.

### 18.6 Use of External Expertise, Advice and Services

A number of the agencies discussed in the previous section would be equally and, in the experience of some businesses, even more at home in this section. But there are a number of other networks of organisations which have the provision of advice and services as their primary activity. This is certainly the case for the occasionally maligned, but certainly prominent, **Business Innovation Centres (BICs)**, which are part of the European Business Innovation Network, and operate to the vague brief of "promoting the growth and development of small-medium firms through innovation". The box below contains a summary of Royce Turner's research (1993) into an attempt to establish a BIC in the heart of a rapidly declining coal field.

## Barnsley Business and Innovation Centre (BBIC)

*As part of a local economic regeneration strategy, BBIC was set up in 1988 with a double objective: to encourage the local economy to diversify away from its dependence on one industry; and to stimulate innovation.*

*The idea behind a BIC is straightforward: innovative products, processes or services can be turned into "market winners", thereby stimulating both local and national economic development and aiding economic modernisation. BICs exist in order to help companies and individuals who might not otherwise be able to develop a particular innovation.*

*The BBIC mission was to "help the growth of new and existing, small-medium enterprises which have innovative technology-based products, processes or services".*

*But, in practice, the BBIC had no hard and fast definition of innovation. Its staff looked for a degree of innovation in the project, but the key factor in a decision to support a project was whether it was replicating something that was already happening locally.*

*Two years down the line, the extent to which the BIC had contributed to the development of businesses which were genuinely innovative was somewhat modest. This was attributed in part to the initial undemanding definition of innovation, but Turner also draws attention to the need for an innovation centre to be operating within a socio-economic environment in which innovation and entrepreneurship are, if not the norm, at least part of the local economic culture. And a mining community, unlike the nearby community in Bradford which contains a large ethnic minority population, has had little reason to develop an "innovative culture".*

*The key lessons, therefore, are clear: innovation cannot be seen simply in local terms; and the creation of an institution needs to go hand in hand with the development of innovation management skills and experience.*

It is not intended that there will ever be a large network of BICs, not least because each one probably needs a catchment area of up to three million people if it is to achieve a real innovative impact. But the *Business Link* network has been established across the whole of England and Scotland, with the objective of stimulating the growth potential of local businesses. Each Business Link has the freedom to develop a range of services to meet local needs, and most house at least one *Innovation and Technology Counsellor (ITC)*, charged to help businesses identify sources of technical assistance, and to aid them in implementing best practice in innovation. The Counsellor receives support from two support mechanisms: *NEARNET* brings together details of all the available local sources of innovation support, while *SUPERNET* is a grouping of technological "Centres of Excellence" either in, or close to, the UK.

## 18.7 Conclusion

This chapter has described the different elements which make up the innovation infrastructure in the UK. Like the previous chapter, it highlights the variety of support available to the business which devotes time and energy to understanding the market in innovation support networks. But a survey of 400 firms with between 50 and 99 employees, undertaken in 1994 by the British Chambers of Commerce, revealed the majority of respondents to be unaware of the SPUR, SMART, LINK and Regional Innovation Grant schemes outlined in this chapter. Yet, despite their imperfect knowledge of what was available "at home", 66 per cent of firms believed that overseas competitors received more assistance than they did with R&D and innovation.

It is clear, therefore, that the majority of small-medium firms will be able to benefit from a greater awareness of the available innovation networks in the UK. Many of the support schemes are tempting, and a few frankly seductive. But they need to be used sparingly and critically, as and where they fit with the strategy of the business (which is why this chapter is towards the end of the book).

They are not an end in themselves, but rather just another aspect of the approaches to networking and partnership which

Part 4 of this book described. And thus their impact depends on there being a dramatic improvement in the climate and local infrastructure in which British businesses operate. The current absence both of collaboration and dynamic business support intermediaries, compared with those examples from elsewhere in Europe that we saw in the previous chapter, has a damaging effect in areas as diverse as training, marketing, technology, R&D, finance and working relationships.

The principal obstacle to the development of thriving industrial districts in the UK remains the isolation of individual small-medium businesses. The debate over whether this is cause or symptom often obscures the fact that not enough is being done to overcome it.

# CONCLUSION:
# INNOVATION AND SOCIETY

Riccardo Petrella[1]

---

*Contents*

---

*This book has examined the nature and processes of innovation in highly developed economies. It contains a summary of current thinking, and both broadens our understanding and sharpens our vision of the present and of the future, as well as raising a number of new questions to which we need to seek answers.*

## 19.1  Innovation Can Take Many Forms — There is no "Best Way"

This book offers a positive and encouraging overview of current thinking in this area. A society can really only move forward when it can accumulate knowledge and experiences which have

---

[1] Riccardo Petrella works for the European Commission within DG XII, Science, Research and Development.

already been tested. Illumination and confirmation have been provided on the following points:

- Unlike the notion of progress — for a long time seen to be a linear and inevitable movement for any society — **innovation is the outcome of a voluntary process.** It depends to a large extent on the existence of a plethora of networks made up of the individuals, groups and institutions who carry projects along, and are capable of defining not only long-term strategies but also the ground rules which can provide a favourable climate for the change process. These incubators for innovation exist sometimes as the initiators of change, sometimes as antennae which are receptive to change occurring elsewhere. Experience certainly proves that those who know how to and are able to innovate are the best equipped to adopt and add value to the innovations of others. This is true not merely for small firms but also for large companies. This is certainly one of the lessons of recent innovation in the field of information technology for the mass market, which has been characterised by a ceaseless interaction between innovations brought about by small firms and then swept up by large companies, which in turn produce a new group of innovations leading to spin offs, and the subsequent creation of new businesses. And all of this takes place within a growing number of alliances and joint ventures between businesses of all sizes in the context of specialist global networks.

- **When it comes to innovation, there is no clear dividing line** between the initial stage of research and development, the subsequent technological innovation, followed by industrial creativity, and then by transfer to the market and finally social innovation. Innovation can occur at any time, at any point at which added value is produced. Nor is there a single vertical line, be it from the top to the bottom or from the bottom to the top. Innovation is the result (as is shown for instance in the success of Minitel in France when compared with the commercial failures of Prestel in the United Kingdom and of Bildschmirtext in Germany) of a good interaction between the two. Finally, there is no compartmentalisation between, on the one side, technological and economic innovation, on the other, social and political innovation and, on the third, cultural innovation. Innovation is a complex whole and its ingredients interact as a result of logic, strategies and conflicts of interest between organised groups and societies, as well as due to similarities and shared values between different communities.

- **Any attempt to search for the perfect model or system of innovation is doomed to failure.** This is one of the key lessons of this book. What work best in practice are those dynamic forms of innovation which turn out to be the most effective at a particular time and in a particular setting. What worked best for Silicon Valley has not been recreated elsewhere, however great the inventiveness devoted to the attempts to imitate its success. In the same way, it is futile to carry out research in "strategic" technologies and in "critical" sectors, with a view to massive investment that it is fervently believed will ensure the capacity for innovation of a business, of a region or of a state. What may seem at one point to be "critical" or "strategic" for a firm or a region may well turn out not to be the case at all for another. Whole economies have managed to develop through investing in sectors which were either in crisis or in decline elsewhere (Japan in the 1960s and 1970s invested heavily in shipbuilding, for example). Today this is equally true for certain sectors which are described as "high-tech". And a majority of the strategies which have been built upon these sectors have led to very disappointing results when compared with the considerable human and financial investment which they absorbed.

- Lastly, **technology is an important and potentially decisive element in the innovation process, but it is only one element, and is not always even the most important**. An obsession with technology is an expensive and frequently disastrous guide, and can lead directly to failure, as is shown in the example of the dissemination (which took place much less rapidly and on a much lesser scale than the "techno-obsessives" predicted) of information technology and automation in manufacturing processes over the last two decades. What is much more likely to have an impact is the right balance between people, technology and the way in which the two are organised. But it should not be forgotten that this balance is never exactly the same from one business to another and from one sector to another, depending as it does on a large number of factors, both internal and external. Each business finds — and must find — its own balance. This will often only be for a short while, since innovation is able to have a rapid impact on the global environment, notably in markets in which the lowering of trade barriers inevitably results in more aggressive competition. In other words, innovation has an everyday impact. It is not possible to live off a single innovation, for the advantages which it brings will be steadily — if not suddenly — eroded.

That was a rapid overview of the current thinking.

### 19.2  Beyond Competitiveness: The Culture of Collaboration

As far as the extension and enriching of present and future visions are concerned, several factors have a bearing on a number of essential areas of which those involved in business, finance and politics are increasingly aware.

The first of these is the low likelihood of transfer of successful examples of innovation and of the patterns of behaviour which characterise them. There are important restrictions on innovation transfer. The intelligent and effective innovator is someone who is able to interpret those principles and key factors in successful innovation which are "lasting"; and is also able to transform this essence in the context of his own specific objectives so as to develop it into the principles and key factors in his own innovation. Outside this action of transformation, there are very few opportunities for transfer to occur. It is becoming increasingly difficult to transfer complicated and sophisticated knowledge and technology, depending as it does on a very rich and diverse innovative environment.

This is one explanation of the ever-widening gap between the most developed economies and the rest of the world. And this gap can only be (more or less rapidly) bridged where the rest have been able to employ their own capacity for transformation.

If new products or services are able to become truly global (as is the case for Levi jeans or Coca-Cola), this will not be because someone has managed to get hold of the very recipe which enabled Levi or Coca-Cola to be successful, but because they will have succeeded in finding *another* recipe which creates new and different products or services.

Secondly, one of the lessons of this book is that to become, and more importantly to *stay*, innovative, an economy or a business needs to develop and work at a ***culture of collaboration***, which both shapes and complements the cult (and culture) of competition which is currently raging throughout most areas of the global economy. The uniqueness and strength of a business is made up of the sum of the human knowledge and skills which it contains, and of the experience and communication which exists within the business and between it and its suppliers, customers and local environment. The culture of competitiveness tends to give undue importance to the present and to the short term, with the prevailing attitude being that if competitive advantage can only be achieved at the cost of eliminating, or even destroying the human, technological, organisational and social experience of other businesses or communities, then so be it. A culture of collaboration, on the other hand, builds innovation upon the synergy and differences which exist among a number of businesses. While businesses which take the logic of conquest to obsessive extremes become, in time, and in everything they do, destructive forces in both economic and social terms.

The culture of collaboration is still more appropriate and coherent when examined in the context of developments which are taking place in knowledge, technology and economics. For instance: the growing cost of R&D and the increasing integration which is taking place between different technologies and areas of understanding; the rising volatility of many markets; the impact and size both of local problems (high rates of unemployment, urban decay) and global issues (such as those to do with the environment, international trade, or stable exchange rates). The culture of collaboration is perhaps most visible at present in the growing number of strategic alliances and joint ventures which have been developed between businesses.

In this context, the role of public bodies is increasingly likely to come to the fore again. This is particularly true when one considers the action which has been taken in some countries to develop scientific and technical culture; or the development of workable and beneficial evaluation systems for public and private sector policy; or the encouragement of public sector/private sector partnerships; or the setting up of new agencies and the passing of new legislation relating to the environment.

What all this demonstrates is that the culture of collaboration is playing an increasingly constructive and healing role, in the teeth of the currently fashionable destructive wave of privatisation, deregulation and economic *laissez aller* with which the last fifteen years are littered.

We have moved on from an era during which we have been subject to the ideology of progress to one in which we bow down before another ideology, that of competitive innovation. Under this banner are to be found a collection of principles, rules, mechanisms and institutions which have turned a single rule guiding the behaviour of economic agents (to flourish in competitive markets) into the sole goal, and the principal objective of public and private-sector players in the economy.

From this point of view, management and business schools have a lot to answer for, since the great majority of them have contributed to the development of the ideology of competitive innovation rather than to the culture of collaboration. In doing this, they have contributed to its debasing and to trivialising the fundamental role that, in every business, is performed by an asset base consisting of the skills and the know-how of its workforce. This experience has been accumulated over time involving interaction and reflection between all levels of the workforce, as well as its roots in the local community, in the local infrastructure and institutional networks, and in the local culture. Many businesses have come painfully to realise this truth, when, dazzled by the lure of greater profit, they have blindly subscribed to the fashion of all-conquering diversification. The majority of these businesses have been able to fend off liquidation by "returning to the knitting" and focusing once more on the area in which they possess a real depth of experience and knowledge.

## 19.3 Signing Up to Global Society

As far as the new issues are concerned, it is worth mentioning here that with the objective of keeping an examination of the nature and application of innovation on solid ground, this book has refrained from entering territory which is both slippery and bumpy. Reading between the lines, however, one finds it suggested on a number of occasions that we need carefully to examine our own behaviour. The reader is left in no doubt of his or her responsibility to rethink the rationale for business activity and the impact of innovation on a business within the broader context of the development of society as a whole and in the light of the most obvious and most pressing problems which affect not only the most developed states but particularly the areas and peoples who are in greatest need.

The question is, on the surface, a simple one: what is the role which innovation (and thus an innovative business) should aspire to play in the future, where it is currently being used to bring about the substitution of technology for human work in production processes, driven solely by the obsessive pursuit of reduced production costs, increased productivity and technological hegemony.

Can we accept that the creation, the development and the diffusion of technology should be the sole source of productive innovation, and that this innovation should be adopted even if it only contributes to increasing, as has been the case in the last 20 years, the level of unemployment in the most developed economies? While at the same time it has only enabled a very small minority of the poor and less developed economies to enter the group of those on an upward development path.

Can we allow ourselves to be satisfied with the argument which goes as follows: "If I don't innovate, unemployment levels will rise even higher", when we know that in most cases competitive innovation not only fails to create more jobs in the innovating business itself, but also invariably succeeds in exporting unemployment to other businesses?

Can we equally be satisfied with the argument which says: "If I don't innovate, the other businesses out there will finish me off. Any economy is about fighting a war to survive. In our market economies, only innovative, aggressive and strong businesses can enter a market and stay there."

These arguments no longer stand up to serious scrutiny. Furthermore, it is unacceptable that we stand by observing the current vogue for financial innovation, which aims only to secure a larger, quicker return than industrial innovation, or innovation which aims to satisfy currently unsatisfied (or only poorly satisfied) social, individual and collective needs.

What we need is a good deal less all-conquering competitive innovation and a good deal more collaborative innovation between businesses, and between businesses and public bodies, in the context of a ***new form of social contract*** on a local, regional, national, continental and even global level.

## 19.4  Innovation and a New Social Contract

A new social contract is indispensable, urgent and necessary, so as to give shape to the enormous potential for innovation which is represented by 420 million Europeans, 1.2 billion Chinese, a billion Indians, 700 million Americans, 550 million Africans, etc.

This new social contract needs to be established on a global scale, following negotiations and agreements at a national and continental level. Its aim should be to give some kind of shape to the two key areas of innovation where businesses have an essential role to play:

- The adoption of new rules governing global economic activity, in an appropriate legislative and political framework

- The development of a number of instruments and measures (that is to say a new mechanism for the world economy) with a view to ensuring that change occurs in practice.

What we are doing at present is passing from a historical period in which the national economy has been the dominant feature, with each state having its own legislative and political bodies, to a new phase: a global economy in which there are fewer and fewer regulatory systems. On one side, we find anarchy (in the case of the financial and monetary system); and on the other side, we find oligarchy (in the case of industrial and financial concentrations of power, and the technological alliances between businesses which are to be found in the global networks of the largest firms). These are the forces which currently regulate and control the world economy. In the same way in which, in the past, the development of national economies was made possible by the regulatory framework which the nation state provided, so the global economy of the end of this millennium can only hope to become an effective base for global development if an equally universal legislative and political framework is established.

Given the small likelihood that a "world state" will come into being in the foreseeable future, it is thus necessary to work to establish a new social contract for effective and concerted innovation, which should cover, at the very least, four elements:

1. Identification of those collective assets that we all share, of which the management and development should respect universal rules of use and access. This concerns, in particular, natural resources, certain basic technical infrastructures and scientific knowledge.

2. Creation of global socio-economic citizenship to accompany the migrations which are inevitably going to increase, if only due to the demographic explosion. This will consist of the progressive promotion of a global policy on work and on population movements. This will use as examples approaches which have been successful elsewhere, such as Erasmus (a European programme to encourage student mobility across the Union), or Fulbright scholarships (enabling researchers and academics to continue their training or to carry out their studies in the United States). Businesses could be given incentives to organise original approaches to the labour market, in the context of public-private partnership projects at a continental, international and global level, and in particular with a North-South perspective. Global social funds could be created, on the model of the European Social Fund. Equally, agreements between Japanese, European and American firms, with a view to trying out new solutions in the areas of employment, work, training, social security and partnership with the South, could be greatly increased.

3. Establishing new global monetary and financial rules. The current lack of order makes it impossible for relations between the different players in an economy to be handled in an efficient way. And innovation is suffering because of this. For where there is the law of the jungle confidence tends not to be very high, and gigantic abuse and corruption (such as, to choose at random from a long list, the failure of the US Savings and Loans, the BCCI scandal, or the laundering of the drug money). It is certain that the problems and solutions are not the same for the developed countries of the North as they are for those of the South.

4. Which brings us to the fourth element: the creation and launch of a number of "solidarity projects", with a view to bridging the gap which has developed between the economies and societies in the North and those in the South of the planet. The dominant economies of this century have secured their progress through, among other factors, vast programmes of collective development: compulsory free education; social housing; major projects involving road construction. Without these, the poor cities and regions of the North would be even more dramatically the victims of their handicaps. And what is true on a national level is even more so on the global level. The first Lomé Convention, signed in 1975 between the European Community and the ex-colonies of Africa, the Caribbean and the Pacific has already demonstrated, despite its limitations, that the rich countries are capable of establishing economic relations with the poor on a more equitable basis, without necessarily having to sacrifice their own interests. So it is not necessary to be afraid of being imaginative in setting up major global projects — and there is no shortage of potential projects: reforestation of the deserts; housing for all; "a billion computers for the Third World"; the list goes on. And why should what has been done, and is still being done, on a national scale (such as the Apollo project in the United States, the high-speed train TGV network in France, or the 120 technology parks in Japan) not be workable on a larger scale?

This new national, continental and international social contract is equally urgently required to serve the interests of businesses. They need to operate within a clear legislative framework, and within transparent guidelines on behaviour. It is not the absence of rules which encourages innovation to flourish and employment to grow, or enables the spread of solidarity and social justice. The absence of rules tends rather to favour aggressive and destructive innovation, which is ever more greedy to devour human skills and knowledge in the same way that it has in the past swallowed natural resources. Innovative businesses are not those which stop at nothing to avoid social or legal responsibility. Recent experience shows clearly that businesses which innovate are those which take responsibility for the social, cultural and environmental costs of their economic activity, rather than in neglecting these responsibilities and attempting to throw them into the lap of society as a whole.

In conclusion, visionary and thoughtful businesses know that they have everything to gain from a new social contract in the place of the global competitive and economic war which it is only in the interest of short-term opportunistic and utilitarian strategies to wage. The time has come for the innovative capacities of businesses and political bodies to turn their backs on this corrupted conception of what a market economy should be.

# GLOSSARY OF KEY TERMS

This section consists of succinct and accessible definitions of around 60 terms.

### Agreements

A formal or informal "contract" between two partners with regard to a transaction. The agreement covers specific issues: price, quantity, quality, timescales. It defines the way in which any eventual dispute will be resolved, and the penalties which will be imposed in case of non-observance of its clauses.

### Appropriate innovation

While the purchase of a new machine or of a patent, or the hiring of a highly qualified technician are major innovations, this does not necessarily make them appropriate innovations for the business. For an innovation to be appropriate, it needs to be integrated into the organisation and into the needs and consciousness of the employees and partners. A new element which is not internalised through active routines runs a high risk of rejection.

### Co-operation (collaboration)

People or businesses working jointly on a project, following the principle of synergy: "The whole is greater than the sum of its parts". Co-operation gives meaning to a technical division of labour within a weak hierarchic framework; it is more egalitarian and stimulates teamwork through a focus on different skills.

### Consultants

Specialists from outside the business, involved in one or other of the functions necessary to its effective operation. These consultants are useful for solving problems or throwing light on particular questions where time is at a premium. They should bring specific skills, without interfering in the life of the business.

## Consultation group

A group of employees from different levels of the organisation who share a specific problem. This group enables the analysis, discussion and confrontation of differing points of view, prior to a decision being taken. It can equally occur at a stage of the process or at the point at which a decision is to be taken.

## Consultation process

The planned economy has occurred at the national level during periods of uncertainty (crises) or of reconstruction (post-war). Social partners (employers, trades unions, government) pool their knowledge, forecasts, and, indirectly, their interests, to influence the decisions to be taken. In the same way, in businesses, the consultation process can precede a decision, and accompany its implementation and the evaluation of the impact. In this way, information which is usually scattered can be shared, points of view can be expressed, and consensus gradually established around a number of shared objectives through shedding light on the decisions to be taken. The process also creates a framework for managing conflict between individual interests.

## Costs

Expenditure to pay for production factors and to cover their implementation: purchases, intermediary consumption, financial and personnel expenses. We can calculate total costs, average and marginal costs, fixed and variable costs, visible and hidden costs. Economic theory originated in an analysis of costs and their minimisation. The organisation of work and economies of scale aim to limit costs. But any industrial strategy must first satisfy a market made up of customers so as to increase market share, quality, flexibility, after-sales service, etc. Here the cost elements are only one of the components of the strategy.

### Critical success factors

A small number of key factors which determine a project's chances of success.

### Databases

Sets of factual data, that are assembled, classified and updated, which concern technological, economic, accounting and financial factors. This data is directly accessible by computer against a rights payment to the owner and data carrier (telecommunications network or the Internet).

### Decision

Choosing between several possible courses of action. Strategic decisions result in general movements whose effects will be seen over a relatively long term, whereas tactical decisions have a more immediate impact. A decision is an isolated act, which has its place within a longer process, beginning with design and ending with implementation and evaluation. It is irreversible.

### Defensive and offensive strategies

Strategies define both the end and the means. Some are responses to circumstances imposed by competitors (defensive), while others aim to destabilise competitors (offensive).

### Diagnosis

Evaluation of strengths and weaknesses of an organisation and its internal processes. The diagnosis consists of several interdependent elements: financial, accounting, organisational, market and technological.

### Effective demand

There are several different theories, but we will focus on the following: the effective demand is that which is considered to exist when the product first reaches the market (even if this demand is not yet visible). Investment decisions, particularly where major innovations are concerned, will be based on this forecast.

### Effectiveness
Results obtained as a consequence of a specific action. Effectiveness of a procedure or of a process. For instance, the increase in the number of products sold as a result of an advertising campaign.

### Efficiency
Synonym for output or productivity. Comparison between the results obtained and the corresponding means employed.

### Environment
Set of external conditions within which an organisation operates. The environment of the business consists both of its direct partners (customers, suppliers, bankers, public services) and of the quality of the infrastructure, legislation, access to technologies. The environment also has an impact on social organisation. It can be related to what happens within the organisation where one considers each function in relation to the others.

### Evaluation
Calculation concerning a particular process. Can be carried out before (chances of success) or after the event (measure of effective results compared to initial objectives and means employed).

### Financial flow
Movement of credits and debits.

### Flexibility
Capacity of an organisation to adapt to change in its environment (partners, markets, technologies).

### Forecasting
Evaluation of possible future developments with regard to a phenomenon. Forecasting may or may not involve numerical calculation. Or it can consist of work undertaken by an expert, the development of scenarios, or econometric models leading to different simulations.

### Hierarchical organisation
A structure where each person occupies a place determined by their function (not necessarily their qualification). It takes the form of a pyramid in which people are classified in ascending levels according to their power.

### Human resources
The sum of people capacity available to a business. Beyond the size of the workforce (permanent, partial and potential) and the spread of qualifications, human resources include know-how, the history of the people and their relationships. Theories of human capital usually reduce human resources to training and health expenditure aimed at increasing productivity.

### Imitation
Reproduction by others of an innovation introduced by a business. The innovating entrepreneur only enjoys a temporary advantage until the arrival of imitators. Imitation can be identical (in which case it will remain a follower) or involve improvements.

### Incremental innovation
Basic innovations whose repetition defines the technological direction. These innovations are initially the result of internal or endogenous behaviours.

### Industrial district
An analysis developed by the British economist A. Marshall and applied most recently in Italian regions such as Prato or Friulia. Describes a dynamic cluster of complementary and competitive activities, usually made up of small-medium firms specialising in certain markets and structured in networks around contractors.

### Information
A set of data on a specific topic. Generic information does not exist; it is always presented either by someone or in a context which gives it meaning. In "information-based companies", an excess is more usual than a lack of information. But one still needs to know how to select the useful information and to give it a meaning which is appropriate to objectives.

### Innovation diffusion
In an exchange economy (not only in a market economy), innovations are spread in a number of ways: by suppliers and customers (any new product "carries" the innovations of which it is made up); by subcontracting (any specification includes innovation at one level or another); by the internal workings of organisations (even the most isolated departments); by memoranda, maintenance notes and know-how. A strategy of innovation diffusion is based on technology watch and on the development of internal information, as well as on more targeted research and development policies. On the other hand, a strategy of innovation protection is based on trade secrets or patents.

### Innovative complementarity
No innovation is developed without there being a knock-on effect on other activities. A new machine will call for organisational innovation or the adaptation of existing machines, of which the improved effectiveness will have repercussions elsewhere, and so on. Innovation is never an isolated act. Optimal results can only be achieved by adapting the organisation, processes and products to each another.

### Investment
In innovation processes, investment takes place as much in kind as in cash. It includes training, the acquisition of the means of production, infrastructure expenses, research and development, the purchase of patents and licenses, and organisation.

### Involving the workforce
Organisation founded on co-operation around shared objectives rather than on the mechanisms of hierarchical authority.

### Joint venture
A company owned by several independent businesses with equal shares. This partnership goes beyond the simple financial dimension: the collaboration can consist of technological or commercial elements, etc.

### Learning by doing

To learn a new technique through practising it. A gradual process, by which one internalises knowledge and reflexes which increase daily efficiency. Simplifies both the acquisition and the development of skills and accelerates production capacities. The learning curve establishes the relationship between productivity gains and experience acquired over time. Beyond the mastering of routines, learning reflects the potential to create new solutions when confronted with production problems; it is also the intersection between the acquisition of internal and external forms of knowledge. The faster the speed of external investment and innovation, the greater the demands made upon the flexibility of the workers who have to move from one technique to another.

### Learning by interacting

Learning through joint work or collaboration. Interaction can come from outside the business, as in the case of a demonstration of new materials; or from the inside when a number of people adopt the best way of using existing equipment or of organising.

### Learning by learning

Initial training through the acquisition of pure knowledge, as a result of the classic process which consists of learning from a teacher who possesses particular knowledge.

### Learning from users

Knowledge derived from observing the behaviour of users of a product and by matching the intrinsic qualities of the product with the needs of the users.

### Life cycle (of businesses, products, technologies or econo-mies in general)

An analogy with biological life cycles based on the observation that fluctuations in economic activity follow recurring patterns. Behaviours, and changes to them, occur in different ways, depending on whether one is in a design phase, a launch phase, a development phase, a phase of maturity, or one of decline.

### Maintenance
Continuous process of servicing existing equipment. This includes tangibles such as machines and other property, or intangibles like skills and organisational systems. The increase in the number of high-technology instruments used in production increases the risk of breakdowns (even though each machine is becoming more and more reliable and sophisticated) and gives maintenance a strategic role.

### Market orientation
Any decision which aims to increase the organisation's ability to take account of market trends or to improve a market position.

### Monitoring the environment
Besides the knowledge and control of the specific conditions of an innovative project (direct cost, technology and the organisation of the appropriate functions), it is necessary to take some general precautions to ensure that the overall conditions in which the company operates can adapt to the specific changes being introduced.

### Negotiation
A transaction between two parties. The life of the business calls for a whole series of explicit or tacit "contracts" between each of its components, thus enabling it to establish some measure of coherence between interests which remain contradictory.

### Networks
Set of relations between different parts of a system. A network is characterised by the nature of its components (the partners who form it and their complementarity), the intensity of relations between them, and the speed of these relations.

### Obsolescence
The outdated nature of a machine, independent of its physical condition. Decline in user value of an object, due to technology, organisational methods or even fashion. Obsolescence concerns not only machines but also organisations, which "wear out" continuously, even if this is not always immediately apparent.

### Opportunity costs
A gain which may result from an alternative use of resources which were allocated to a particular expense.

### Organisation
Structuring of interpersonal relations and rules of behaviour within institutions.

### Partnership
Capacity to get individuals with complementary skills and different interests to work together on the same project.

### Patents
The right conferred by public authorities to the inventor of a product or a process to enjoy a temporary monopoly for its exploitation. The patent is only delivered after proof that it is genuinely new. It can be kept and exploited by its owner, sold or licensed for use in specified conditions.

### Performance indicators
A measure of final results obtained: volume of stocks, time to set up machine, number of breakdowns, etc.

### Process innovation
Improvement of the tangible (machines) and intangible (skills, software) means of production. Any process innovation involves organisational innovation.

### Product innovation
Creation of a new product or improvement of an existing product. Innovation directly relating to the market (whether or not it already exists). Always involves process and organisational innovation.

### Productivity
Relationship between an output and what was used or spent to obtain it. It is possible to calculate the productivity of people or machines. The "apparent" productivity of work is the relationship between the added value and the amount of work, represented by

the number of hours spent. However, an improvement in production, all factors being constant, can have many causes which need not be exclusive: intensity of work, economies of scale, experience, organisation, modernisation of equipment and incorporation of technical progress.

### Product life cycle

Theory developed by the American R. Vernon which relates the development stage of a product to international business strategies. Initially, the product cannot appear in the most advanced areas; competition operates on criteria of innovation and suitability to potential demand. International development is only possible in the growth phase as one attempts to standardise the product. In the maturity phase, the product is accessible to all and the technology is standardised. Competition then operates on price and production costs.

### Quality

Qualitative characteristics of a product. The quality/price ratio of a product or service put on the market is one of the elements of competitiveness. Against the theory, the quality of products is never homogenous.

### Radical innovation

Changes of a technological kind, whether in the economy as a whole or in a given business. The steam engine or microelectronics are radical innovations for a society; the transition to just-in-time production is a radical innovation for a business at a certain point. These innovations are mainly "imported", of external origin or exogenous.

### Resistance to change

Any institution, like any individual, is even more terrified of change where they consider themselves to have invested heavily to acquire what they currently have. Resistance to change is a normal tendency, and needs to be taken into account in any business strategy.

### Risk
Giving material form to the unpredictable nature of the future. Uncertainty, applied to projects and corresponding commitments, enables the calculation of levels of risk, ranging from near certainty to a total (and incalculable) risk. Risk cannot be reduced to its probability; it also includes the relative scale of the stakes, and the size of the event: level of potential financial losses, of the loss of market share, employee demotivation, take-over by a competitor, etc. Behaviours regarding risk are highly dependant on psychological elements (risk aversion, risk takers).

### Service centres
Public or public-private organisations which can be local or regional, equipped to provide advice or direct services to small-medium firms: business incubators are service centres, as are industrial technical centres (ceramics, electronics, textiles...), training centres. Services are usually sold, either at their true cost, or on the basis of shared costs.

### Strategy
Set of co-ordinated actions to meet long-term objectives for the business or for the country. A military concept originally, strategy is the reflection of the relationship between activity on the ground and objectives pursued. National technological strategies cover the whole "technological policy"; those at the level of the business can include access to technology, the means to develop all forms of innovation, and the means chosen to add value to the results or the markets for goods or patents.

### Synergy
Combination of two or more converging factors which have a final result which is superior to the mere addition of the initial factors.

### Tactics
Set of co-ordinated means used to achieve a result.

### Technological assets
All the knowledge, tools and routines acquired by the business. These assets can be recognised, explicit and physical (patents), or

remain entirely implicit, even secret, and be incorporated into the skills of the employees and their behaviours.

### Technology transfer

Way in which an innovation circulates from its creator and its initial use to other users and other uses. We usually talk of technological transfers between developed and under-developed countries; but between businesses within developed countries we talk more of diffusion of technologies or diffusion of innovation.

### Technology watch

Evaluation of potential developments outside the business concerning available technologies and client strategies in this area, as well as the different dimensions of the environment. Watching does not include only the gathering of information but also its process and diffusion.

### Training

One of the structured ways of acquiring knowledge. Direct, explicit and intentional transfer of generic or specific knowledge.

### Transaction costs

Two meanings. On the one hand, associated costs related to buying and selling: information costs, negotiation time, possible legal discussions and disagreements, etc. On the other hand, the expenses linked to the organisation of industrial relations outside market operations: time for salary negotiations, organisation of the smooth running of services, efforts made for sub-contracting relations to be more productive, communication with customers and suppliers outside specific orders, etc.

### Uncertainty

Unpredictability of the future. Uncertainty can be virtually zero (the business will pay on time) or total (launch of an entirely new product). It can be calculable (automobile risks for an insurance company) or incalculable (explosion of a nuclear reactor, a space rocket or the theft of the Mona Lisa from the Louvre museum).

# BIBLIOGRAPHY
(of key texts on the central themes of this book)

Ansoff, Igor (1965), *Corporate Strategy* (London: Penguin)

Bagnasco, A. and Sabel, Charles (1994), *PME et Développement Economique en Europe* (Paris: La Découverte)

Barrow, Colin et al. (1993), *Growing Up* (London: etc limited)

Bommensath, M. (1991), *Secrets de Réussite de l'Éntreprise Allemande* (Paris: Editions d'Organisation)

Burns, T. and Stalker G. (1961), *The Management of Innovation* (Oxford: Oxford University Press)

Butera, Federico (1990), *Il Castello et la Rete* (Milan: Angeli)

Buxton, T. et al. (1994), *Britain's Economic Performance* (London: Routledge)

Crozier, M. and Friedberg, E. (1977), *L'Acteur et le Système* (Paris: Le Seuil)

Davenport, Thomas H. (1993), *Process Innovation* (Cambridge, MA: Harvard Business School Press)

Dixon, Norman (1994), *On the Psychology of Military Incompetence* (London: Pimlico)

Dosi, G. ed. (1984), *Technical Change and Industrial Transformation* (London: Macmillan)

Doyle, P., Saunders J. and Wong, V. (1986), "Japanese Marketing Strategies in the UK", *Journal of International Business Studies*, Vol. 17, No.1

Drucker, Peter F. (1985), *Innovation and Entrepreneurship* (London: Pan)

European Commission (1995), *Green Paper on Innovation* (Brussels: European Commission)

European Innovation Programme (1995), *The European Handbook of Management Consultancy* (Dublin: Oak Tree Press)

Formaper (1992), *Innovazione Tecnologica e Bisogni Formativi in un Campione di PMI dell Area Milanese* (Milan)

Freeman, Christopher (1982), *The Economics of Industrial Innovation* (London: Pinter)

Geroski, P. and Machin, S. (1992), "Do innovating firms outperform non-innovators?" (*Business Strategy Review*, Summer)

Grant, Robert M. (1991), *Contemporary Strategy Analysis* (Oxford: Blackwell)

Handy, Charles (1976), *Understanding Organisations* (London: Penguin)

Hassink, R. (1992), *Regional Innovation Policy* (Utrecht)

Heart of England TEC (1994), *Innovation Action Tool Kit* (Oxford: Heart of England TEC)

Hendry, C. et al. (1995), *Strategy through People* (London: Routledge)

Hutton, W. (1995), *The State We're In* (London: Jonathan Cape)

Kanter, Rosabeth Moss (1992), *The Change Masters* (London: Routledge)

Lasfargues, Yves (1988), *Technojolies, Technofolies* (Paris: Editions d'Organisation)

Massey, D. and Wield, D. (1992), "Evaluating Science Parks", *Local Economy*, Vol. 7, No. 1

Ors Valle, J. (1994), "Small Firm Innovation Networks in the Valencia Region", *European Planning Studies*, Vol. 2, No. 2

Pettigrew, A. (1973), *The Politics of Organisational Decision-making* (London: Tavistock)

Piore, M. and Sabel, C. (1984), *The Second Industrial Divide* (New York: Basic Books)

Porter, Michael (1985), *Competitive Advantage* (New York: The Free Press)

Riboud, A. (1987), *Modernisation, mode d'emploi* (Paris: 10/18)

Rosenberg, N. (1982), *Inside the Black Box, Technology and Economics* (Cambridge: Cambridge University Press)

Schein, E. (1988), *Process Consultation* (Reading, MA: Addison-Wesley)

Schumpeter, J. (1926), *Theory of Economic Development* (London: George Allen and Unwin)

Senge, P. (1994), *The Fifth Discipline Fieldbook: Strategies and Tools for Building a Learning Organisation* (London: Nicholas Brealey)

Stanworth, J. and Gray, C. eds. (1991), *Bolton 20 Years On — The Small Firm in the 1990s* (London: Paul Chapman)

Steele, Lowell W. (1988), *Managing Technology* (New York: McGraw-Hill)

Storey, D.J. (1994), *Understanding the Small Business Sector* (London: Routledge)

Teece, D. et al. (1988), *Technical Change and Economic Theory* (London: Pinter)

Turner, Royce (1993), "From Coal Mining to a High Technology Economy?", *New Technology, Work and Employment*, Vol. 8, No. 1

West London TEC (1994), *Economic Development Workbook* (London: West London TEC)

Wilson, David C. (1971), *A Strategy of Change* (London: Routledge)

Wilson, David C. and Rosenfeld, R. (1990), *Managing Organisations* (Maidenhead: McGraw-Hill)

# INDEX